ORGANISMS AND ENVIRONMENTS
Harry W. Greene, Consulting Editor

1. *The View from Bald Hill: Thirty Years in an Arizona Grassland,* by Carl E. Bock and Jane H. Bock
2. *Tupai: A Field Study of Bornean Treeshrews,* by Louise H. Emmons
3. *Singing the Turtles to Sea: The Comcáac (Seri) Art and Science of Reptiles,* by Gary Paul Nabhan
4. *Amphibians and Reptiles of Baja California, Its Pacific Islands, and the Islands in the Sea of Cortés,* by L. Lee Grismer

Tupai

*The publisher gratefully acknowledges
the generous contribution to this book provided by
the General Endowment Fund of the Associates
of the University of California Press.*

Tupai

*A Field Study
of Bornean Treeshrews*

Louise H. Emmons

Foreword by Harry W. Greene

UNIVERSITY OF CALIFORNIA PRESS
Berkeley · Los Angeles · London

University of California Press
Berkeley and Los Angeles, California

University of California Press, Ltd.
London, England

© 2000 by the Regents of the University
of California

Library of Congress Cataloging-in-Publication Data

Emmons, Louise.
    Tupai : a field study of Bornean treeshrews /
Louise H. Emmons ; with a foreword by Harry W.
Greene.
       p.   cm.—(Organisms and environments ; 2)
    Includes bibliographical references and index.
    ISBN 0-520-22291-1 (alk. paper).—
    ISBN 0-520-22384-5 (pbk. : alk paper)
    1. Tupaiidae. 2. Tupaiidae—Borneo.
I. Title. II. Series
QL737.S254 E44   2000
599.33'8'095983—dc21   99-089484

Manufactured in the United States of America
09 08 07 06 05 04 03 02 01 00
10 9 8 7 6 5 4 3 2 1

The paper used in this publication meets the minimum requirements of ANSI/NISO Z39.48-1992
(R 1997) (*Permanence of Paper*).

Contents

List of Illustrations	vii
List of Tables	xi
Foreword	xiii
Acknowledgments	xvii
1. *Tupai:* An Introduction	1
2. The Study Species	7
3. The Milieu: Field Study Sites and Habitats	24
4. Treeshrews in Their Habitat	38
5. Diet and Foraging Behavior	53
6. Nesting Behavior	91
7. Activity Patterns	110
8. Use of Space	124
9. Social Organization	145
10. Life History	169
11. Predation, Predators, and Alarm Behaviors	202
12. Synthesis	210
Appendix I. Methods	227
Appendix II. Fruit Species Collected at Danum Valley	232
Appendix III. Mammal Species Found on the Study Plots	237
Appendix IV. Invertebrates in Treeshrew Diets	240
Appendix V. Consumers of Fruit Species	244
Appendix VI. Response of Murid Rodents to the Masting Phenomenon of 1990–1991	247
Bibliography	251
Index	261

Illustrations

1.1.	A portrait of a *Tupaia tana*.	2
2.1.	The Jurassic pantothere *Crusafontia* and a large treeshrew, *Tupaia tana*.	9
2.2.	Skulls of large treeshrew, *Tupaia tana*, and lesser treeshrew, *Tupaia minor*.	11
2.3.	Portraits of *Ptilocercus lowii*, *Tupaia longipes*, *Tupaia minor*, and *Tupaia gracilis*.	18
2.4.	Mean body mass of the treeshrew study species.	21
3.1.	Location of study sites.	25
3.2.	Dipterocarp tree near Poring.	26
3.3.	Map of trails and trap sites on the lower plot at Poring.	28
3.4.	The forest at 600 m on the study plot at Poring.	29
3.5.	The forest at about 900 m on the high study plot at Poring, habitat of *T. montana*.	30
3.6.	Helicopter view of Danum Valley Field Centre.	31
3.7.	Map of the trails and streams on the study plot at Danum Valley.	32
3.8.	Helicopter view of the forest canopy on the study plot at Danum Valley.	34
3.9.	Monthly rainfall and number of fruit species collected per month during the study period at Danum Valley.	35
4.1.	Height distributions of treeshrews in the two study areas.	39
5.1.	Influence of fruit trees on treeshrew movements.	54

5.2.	Size of fruits eaten by *Tupaia* species and size of fruits collected in transects at Danum Valley.	57
5.3.	Fruit of Sapotaceae (cf. *Payena*) and wads of fiber spit out by *T. longipes*.	60
5.4.	Fruit species and total weight of fruit collected monthly on the Danum Valley study area per 100 m of trail, and rainfall during those months.	63
5.5.	The percentage of treeshrew scats (all species) collected on the Danum study area that contained fruit, by month.	65
5.6.	Percent occurence of food items in treeshrew diets.	70
5.7.	Average numbers of arthropods in samples of litter collected at Danum Valley and total numbers of arthropods collected from ten Bornean trees by canopy fogging.	77
5.8.	Numbers of litter invertebrates collected in litter samples during three-month periods in 1988–89.	78
5.9.	Distribution of sightings of bulbuls and drongos following *T. minor* at Danum Valley.	89
6.1.	The base of the pentail nest tree.	93
6.2.	The entrance, almost undetectable, of a *T. longipes* burrow.	95
6.3.	A nest constructed only of overlapping leaves from a burrow of *T. longipes*.	96
6.4.	Nest site of a *T. montana* in a ground cavity.	97
6.5.	A nest of *T. montana*, constructed only of woody fibers.	98
6.6.	The tree hole that contained the nest with young of *T. tana* F109.	99
6.7.	An exposed leaf nest of *T. tana* F109.	100
7.1.	Times at which treeshrews entered their nests at the end of activity.	112
7.2.	Distribution of inactivity recorded during radio-tracking.	115
7.3.	Number of rainy days per month at Danum Valley in 1990 and 1991.	116
7.4.	Hour of onset of rain at Danum Valley, all recorded showers.	117
7.5.	Number of hours during which rain fell during each month.	118
8.1.	Number of radio-tracking location points recorded for each treeshrew followed and its home range in hectares.	128
8.2.	Treeshrew home range areas as a function of body mass.	129

8.3.	The relationship between meters traveled per day and area of home range, for all individuals of each species.	132
8.4.	Contrasting patterns of daily use of space.	133
8.5.	Mean distance traveled per day by treeshrew species.	133
8.6.	The general relationship between mean meters traveled per day and area of home range; data for all treeshrews.	134
8.7.	The distance traveled during a day's activity as a function of the maximum diameter of the range for that day.	135
8.8.	Two sequential one-day itineraries of *T. tana* F100 that show use of stream courses.	139
9.1.	Home ranges of the four pentail treeshrews.	148
9.2.	Territories of *T. minor*.	149
9.3.	Ranges of the three *T. gracilis* followed at Danum Valley.	152
9.4.	Territories of neighboring pairs of *T. longipes* and the range of a subadult male at Danum Valley.	154
9.5.	Takeover in March 1991 of half of the territory of *T. longipes* female F56 by female F86.	155
9.6.	Territories of *T. montana* followed on the high plot at Poring.	157
9.7.	Territories of nine adult *T. tana* at Danum Valley from September to December 1990.	158
9.8.	The arrangement of territories of *T. tana* from March to September 1991.	159
10.1.	Nestling *T. tana*, 26 April 1989.	172
10.2.	Seasonal reproductive activity of female treeshrews trapped at Danum Valley.	179
10.3.	Seasonality of the appearance of young on the study area at Danum Valley.	180
10.4.	The month of first capture of young of *Tupaia* species at Danum Valley; fruit species collected monthly and percent of treeshrew scats including fruit; monthly rainfall; and numbers of litter invertebrates collected.	183
10.5.	Histories of all individuals of *T. minor* and *T. gracilis* captured on the Danum Valley study area.	185
10.6.	Histories of individual *T. longipes* trapped at Danum Valley.	186
10.7.	Histories of individual *T. tana* on the study site at Danum Valley.	187

10.8.	Number of treeshrews captured and number known alive each month at Danum Valley.	191
12.1.	Schematic diagram of the axes of ecological separation between the six treeshrew species studied.	211
A-I.1.	Total number of treeshrews trapped monthly at Danum Valley, September 1990 to September 1991.	228
A-VI.1.	Murid rodents captured monthly at Danum Valley.	248

Tables

2.1.	Mean length of intestinal segments in treeshrews and some other mammals.	13
2.2.	Treeshrew species studied, standard measurements of males and females.	20
4.1.	Distribution of sightings of treeshrews on various substrates.	44
4.2.	Chief direction of movement of treeshrews observed foraging arboreally.	44
4.3.	Elevational limits observed in *Tupaia* species on Mount Kinabalu.	47
4.4.	Comparison of captures on unlogged study plot and selectively logged forest at Danum Valley, and results of trapping in logged forest and tree plantations by Stuebing and Gasis (1989).	49
5.1.	Fruit species eaten by treeshrews.	55
5.2.	Fruit transit time for two species of *Tupaia* in captivity.	59
5.3.	Time spent by females of *T. tana* (F65), *T. longipes* (F56), and *T. gracilis* (F67) at a fruiting mata kuching (*Dimocarpus longan*).	62
5.4.	Months during which fruit species were eaten by treeshrews at Danum Valley.	64
5.5.	Occurrence of food items in diet.	71
5.6.	The five major invertebrate taxa eaten by each treeshrew species.	72

5.7.	Summary of invertebrate foraging characteristics of the six treeshrew species.	80
5.8.	Foraging behaviors of some babblers at Danum Valley and the treeshrews they are likely to compete with for prey.	86
5.9.	Associations of *Tupaia minor* with birds.	88
6.1.	Dimensions of burrows and included nests of *T. longipes* at Poring.	94
6.2.	Numbers of sleeping sites used by treeshrews during radio-tracking.	102
6.3.	Summary of nest site locations and nest constructions.	104
7.1.	Activity statistics of treeshrews.	113
8.1.	Home range data for 48 treeshrews.	126
8.2.	Home range data, means for species and sexes.	127
8.3.	Mean distances traveled by treeshrews per day.	130
8.4.	The relationship between time spent at fruit trees during a day, the number of visits to the tree, and the distance traveled per day by treeshrews.	137
8.5.	Estimated treeshrew population densities.	141
8.6.	Ranging behavior of other small, insectivorous/frugivorous equatorial rainforest taxa.	143
9.1.	Groupings of treeshrews.	150
10.1.	The behavior of *T. tana* male M111 at Poring after the birth of F109's young.	174
10.2.	Litter sizes of the *Tupaia* study species.	176
10.3.	Reproductive histories of resident female treeshrews at Danum Valley, 1990–91.	178
10.4.	Reproductive histories of female treeshrews at Poring Hotsprings, 1989.	182
10.5.	Schema of possible functions of absentee maternal care and the probable "reason" for the system in other mammals.	197
11.1.	Number of recorded sightings of mammalian predators in the forest on or near the study areas.	206
12.1.	Summary of ranging patterns of females and foraging characteristics of treeshrew species.	212

Foreword

Harry W. Greene

Tupai: A Field Study of Bornean Treeshrews is the second volume in the University of California Press's new series on organisms and environments. Our main themes are the diversity of plants and animals, the ways in which they interact with each other and with their surroundings, and the broader implications of those relationships for science and society. We seek books that promote unusual, even unexpected connections among seemingly disparate topics; we want to encourage writing that is special by virtue of the unique perspectives and talents of its authors. Our first installment, by Carl E. Bock and Jane H. Bock, synthesizes thirty years of experimental research on Arizona grasslands, and an ethnoherpetology of the Seri by Gary Paul Nabhan is forthcoming.

Louise H. Emmons's *Tupai* reports on an in-depth field study of small, poorly known, and as it turns out very peculiar animals—and therein lies an important message. Sometimes lost in the excitement of modern life science, with its technological wizardry and emphasis on human health, is the fact that functioning organisms themselves are still a central concern of biology; their very diversity is the big mystery. Moreover, most species have rarely, if ever, been observed, and among them are some so unusual that they might recast our theoretical frameworks, enlarge the boundaries of what we seek to explain. Where do these unstudied creatures drink and sleep, and what are their daily transactions with predators, prey, and relatives? Do their distinctive characteristics facilitate life in particular environments, and how will they respond to ever-increasing human impact? Answering those questions, even for large species in open habitats, requires patience, ingenuity, and stamina; field studies also entail supplies, research permits, and funding. Describing in detail the lifestyles of animals that are secretive, shy, nocturnal, and living in dense vegetation demands all that and something more, a certain blind faith in the task, an uncommon reservoir of curiosity and devotion to studying nature.

Given the many inherent difficulties, what kind of person would set out to watch treeshrews in Borneo? Louise Emmons is widely admired among vertebrate biologists for her breadth of interest and experience, ability to work under exceptionally remote and rugged conditions, and scholarly attention to details. Since earning a Ph.D. at Cornell University for her comparative study of West African rainforest squirrels, Louise has ranged all over the world and studied a remarkable array of organisms. Now a research associate at the Smithsonian Institution, she has written more than sixty scientific publications concerning everything from the giant flowers of *Rafflesia* and tortoise predation by jaguars to radiotelemetry of brush-tailed porcupines and ocelots. Along the way she discovered two new genera and several new species of rodents and documented the distribution and natural history of many other poorly known mammals. Louise has served on rapid assessment teams for Conservation International, surveying potential park sites and training local naturalists in some of the remotest parts of South America, and has written a highly acclaimed field guide to Neotropical mammals, now available in Spanish.

Treeshrews, known as *tupai* in Malaysia and Indonesia, are restricted to the once-vast tropical forests of southern Asia and nearby islands. They live with clouded leopards, monitor lizards, and other stealthy predators, and indeed treeshrews behave as if danger lurks behind every fallen log, as if each root in their path is a snake. Once heralded as primitive "proto-primates," they actually are neither shrews nor close relatives of monkeys. Superficially squirrel-like in appearance, treeshrews have oddly shaped noses, surprisingly rich milk, and diverse social systems; the absentee maternal behavior of some species is so bizarre that it cannot have been ancestral to anything else known among living mammals. Until Louise's eighteen-month study, most of what we knew of these biologically interesting creatures came from laboratory studies and isolated field observations. Her descriptions reveal a fascinating mix of relatively primitive and highly derived, sometimes unique characteristics and thereby provide a basis for comparisons with close relatives and other mammals. The variation she uncovered among species not only yields important insights for more broadly understanding the evolution of foraging and social behavior but also predicts the differential vulnerability of those treeshrews to ongoing deforestation.

Tupai is a vivid, richly textured account of original research, and anyone interested in the ecology, behavior, and evolution of mammals will profit from this fine book. I especially recommend it for new graduate students and others contemplating difficult fieldwork, all the more so because Louise is candid about frustrations and setbacks. And for everyone enthralled by nature, *Tupai* enriches our appreciation for the variety of life, illuminates the career of a superb biologist, and inspires us to ponder what it is like to be a treeshrew.

Acknowledgments

The fieldwork that is described in this book was done between 1989 and 1991 in Borneo. Although I worked as an independent researcher, the study would have been much diminished or could never have been done without the support of many institutions and the help of many people, both old friends and new ones whom I found or who found me along the way. The expeditions were most generously financed by the Douroucouli Foundation, the Chicago Zoological Society, Brookfield Zoo, and the National Geographic Society. The Smithsonian Institution Division of Mammals housed me and provided office facilities throughout this and many other projects. The Universiti Kebangsaan Malaysia Kampus Sabah (UKMS) and its dean at that time, Dr. Ghazally Ismail, welcomed me to their country and campus and helped to facilitate my research in many ways. My work in Mount Kinabalu Park was encouraged and supported by Lamri Alisaputra and Francis Liew of the Sabah Parks Department. My stay in the park was enhanced by the warm friendship and hospitality of Gabriel Sinit, Tan Fui Lan, and Jamili Nias. For their unstinting and uncomplaining assistance in the field at Poring, during conditions that were often difficult, I cannot thank enough Hajinin Hussain and Alim Biun, who also taught me much about the forest and the Dusun people. For making it possible for me to work at Danum Valley Field Centre, I am indebted to Dr. Clive Marsh. Mohammad Dzuhari gave able assistance in the field at Danum. I could not have undertaken the fruit collection project without the expert botanical help of Elaine Gasis.

Han Kwai Hin transcribed data for me from specimens in the UKMS Museum; Dr. John Cadle identified a snake; Stewart Diekmeyer helped with the photographs; and Dr. Fred Sheldon found breeding bird records for me among his data from Sabah. Warren Steiner of the Smithsonian Institution undertook the Herculean task of identifying insect remains from scats; his care and expertise produced extraordinary results. For reading all or part of the manuscript and improving it with critical comments, I thank Frank Lambert, Patricia Wright, Larry Heaney, David Woodruff, and Christy Rabke. Miles Roberts freely shared his results on captive treeshrews, and I regret that his fine work could not be included in this volume. To Harry Greene I owe a special debt for moral support and encouraging me to submit the manuscript to the University of California Press. For their help and companionship and stimulating conversations in the field or in Kota Kinabalu, I am grateful to Frank Lambert, Sharon Emerson, Lindai Lee, Joseph Gasis, Christine Eckstrom, Frans Lanting, Michelle Pinard, Junaidi Payne, Charles Francis, and Steven Pinfield. I also thank Frans Lanting for his splendid cover photographs. Finally, for welcoming me into their home in Kota Kinabalu as a member of their family, and for their time, help, and support in innumerable small and large tasks, I thank Robert and Ping Stuebing, whose friendship made all my stays in KK a great pleasure.

CHAPTER 1
Tupai: An Introduction

Treeshrews[1] suffer from chronic mistaken identity: first, they are not shrews; second, most are not found in trees; and third, what they really are (among mammalian orders) has never been agreed on. The first treeshrew recorded by a Western naturalist was collected in 1780 by William Ellis, surgeon to Captain Cook's expedition (Lyon 1913). He thought it was a squirrel. Indeed, at first glance, treeshrews are so much like squirrels that people who live where they occur often confuse the two, and they are known by the same common name, *tupai,* in the Malay/Indonesian language. This local name was the basis for the first treeshrew generic name, *Tupaia,* and its family, Tupaiidae. Forty years passed before taxonomists recognized treeshrews as different from squirrels (Lyon 1913). These animals, which would better have been called squirrelshrews, are in fact like no others.

The best-known treeshrews, in the genus *Tupaia,* are active, alert creatures that border on the neurotic. They are squirrel-sized, brownish mammals with large, dark, lashless eyes; short, bare ears; and a large, wet nosepad like a pencil eraser on the tip of a long muzzle (fig. 1.1). They have squirrel-like feet; soft, dense fur; inconspicuous whiskers; and a long bushy tail that flicks upward. When alarmed, they chatter, whine, or whis-

---

1. I use the single word *treeshrew,* rather than *tree shrew* or *tree-shrew,* to distinguish these animals as clearly as possible from genuine shrews of the order Insectivora (long-tailed shrew, pygmy shrew, etc.).

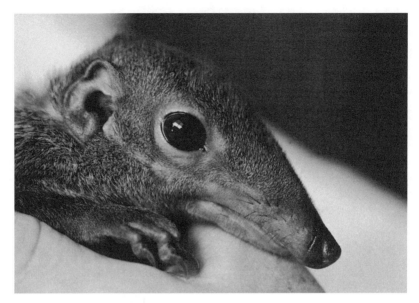

Fig. 1.1. A portrait of a *Tupaia tana*.

tle. At a sudden noise, they flinch as if struck. It is hard to escape their notice, and wary species are difficult to watch for more than a few moments. One usually catches a brief glimpse of a small brown animal flashing across a trail or racing down a log. Treeshrews bound with springing leaps from sapling to sapling, ferret slowly along the ground, or rustle through the foliage of a tree. Close up, there is something alien about treeshrews, with the ancient expression in their wide eyes and their long down-turned mouths.

WHAT IS A TREESHREW?

The question, What is a treeshrew? has been asked many times, almost always while seeking the answer to the treeshrew's place in evolutionary history. In this book I am mainly concerned with other questions, but because a living animal is a result of its history, I begin with a brief overview of thought on treeshrew taxonomy and phylogeny—the search for their true, but often mistaken, identity.

Before 1900 treeshrews were generally thought to be in the order Insectivora and related to the true shrews, but by the first decade of the twentieth century comparative anatomists separated them from that order and proposed a closer relationship to other groups, especially ele-

Introduction

phant shrews (Macroscelididae; e.g., Gregory 1910). In 1922 Carlsson placed them in the order Primates, and this was subsequently supported with monumental anatomical studies by Le Gros Clark (1925, 1959; see also Elliot 1971). This view dominated for the next half century (e.g., Simpson 1945), stimulating mountains of comparative work on morphology and physiology that was strongly focused on the paradigm of treeshrews as basal primates (although there was never total consensus on their placement in that order). In 1980 a fresh outlook was brought by the application of modern methods of analysis for phylogenetic reconstruction. In a comprehensive, multiauthor review, nearly every well-studied set of features of treeshrew anatomy was compared to those of possibly related taxa of other orders (Luckett 1980). The editor, W. P. Luckett, concluded, "Following evaluations of the available dental, cranial, postcranial, neuroanatomical, reproductive, developmental, and molecular evidence, the contributors were in general agreement that tupaiids do not appear to share any uniquely derived features with living or fossil primates that would warrant inclusion of tree shrews in a monophyletic order Primates" (1980: viii). Calling treeshrews primates was another case of mistaken identity.

This has returned us to perspectives similar to those of 1910 and earlier. Several hypotheses, one more than a century old, have grouped treeshrews with other mammalian orders, primarily Macroscelidea, Dermoptera (flying lemurs), and Lagomorpha (rabbits), and into a superorder with these and/or primates and bats, but there is still no consensus. Clearly, the last ancestor that tupaiids shared with members of other living orders was so long ago that no unambiguous set of morphological characters allies treeshrews more closely with one order than with another. Along with most other mammalogists, I consider treeshrews to belong in their own order, Scandentia Wagner, 1855, and await more robust evidence of their deeper systematic relationships. Treeshrews are treeshrews.

The only complete review of the genera and species of treeshrews was the classic work of Lyon (1913), who reviewed all available forms, named many new ones, and in the process discovered many important physical features (taxonomic characters) useful for classification. Lyon recognized six genera and forty-six species of treeshrews. Since then, many of these have been grouped together as synonyms, especially island populations of the genus *Tupaia*, to the point that only five genera and sixteen species were recognized (Corbet and Hill 1991, 1992; Honacki, Kinman, and Koeppl 1982). However, Wilson (1993) recently reversed this trend by recognizing nineteen species. I follow Wilson's classification, which is pertinent to my work because it gives species status to *Tupaia longipes,* a treeshrew I studied, and distinguishes this Bornean population from the West Malaysian *T. glis.* This distinction seems to be supported by our data. A modern revision of all treeshrews is overdue, and when this is done a number of other forms now buried in synonymy are likely to be resurrected to species rank. However, the species whose ecology and behavior I describe here are distinct biological entities, and changes in future taxonomies will not invalidate or obscure the subjects of this work. The order Scandentia is endemic to the Indomalayan region and the Philippines, and Borneo, with ten species, is overwhelmingly the center of living treeshrew diversity (Corbet and Hill 1992). To Borneo, then, I went to study them, because only in Borneo could the full range of treeshrews be found together.

WHY STUDY TREESHREWS?

As representatives of one of only twenty-six living orders of mammals, treeshrews are worth studying just to discover the characteristics of a major branch of the Mammalia, but aside from pure curiosity, there are other reasons to find out more about these obscure small animals. Although treeshrews are no longer considered primates, they are thought by some to belong to a group of closely related orders (a grandorder), the Archonta, which besides treeshrews includes the primates, bats, and flying lemurs. In this group treeshrews are the least specialized (most primitive) members. As such, they may still be the most closely related living models of the very earliest primate ancestors of the late Cretaceous period. Their lifestyles can thus provide a window on our earliest antecedents, and perhaps a view of why evolution may have taken the direction it did.

Almost all of the 220 million years of mammalian history are recorded only by fossil bones, more often than not from mere fragments, a teaspoonful of teeth. For the first two-thirds of mammalian history, all mammals were small to tiny carnivores or insectivores (Lillegraven, Kielan-Jaworowska, and Clemens 1979). From these inconspicuous creatures came the great Paleocene evolutionary radiations of mammals, which followed the extinction of the dinosaurs at the end of the Cretaceous period 65 million years ago. We can only guess how these early mammals might have used their habitats. Such guesses are largely built from analogies drawn from living species that have similar morphology: we assume that the saber-toothed marsupial *Thylacosmilus* was a carnivore like the saber-toothed cats; and that the strange South American litoptern ungulate *Macrauchenia* had a trunk and browsed on leaves, because its skull has a single nostril between the eyes, as do elephants today (Benton 1991). Morphology can sketch only the faintest outline of how an animal really lives. It may suggest that a species eats insects; but what kind of insects? Where does it find them? Does it feed by day, or night, or both? Does it live alone, have a territory, or share its home with others? Treeshrews have many morphological similarities to early, or primitive, mammals. One of the reasons to study them is to find out what such animals do and what they are capable of. Are different species much alike in how they use their environment? Can we predict the details of their lifestyle from their structure? What, ecologically speaking, is a primitive mammal? Is there such a thing? Can we infer anything about mammalian evolution from the behavior of living species? By studying how treeshrews behave and live in their natural habitats, we can perhaps start to answer such questions.

EARLIER STUDIES

During the half century of treeshrew glory, when treeshrews were considered primates, enormous quantities of research were done on their anatomy and physiology, and also some behavioral work on laboratory colonies. A bibliography of treeshrews published in 1971 contains 1,036 references (Elliot 1971), almost all laboratory or zoo studies of captive-bred animals. The subjects of almost all of this work were laboratory stock purportedly of a single species (but perhaps three), *Tupaia glis* and its related forms *belangeri* and *chinensis*, whose origins were usually unknown but who were thought to come from Thailand. These became a mythical entity: "The Treeshrew."

In a pioneering study at Seewiesen, Germany, where Konrad Lorenz had introduced entirely new approaches to research on animal behavior, Robert D. Martin (1968) conducted the first comprehensive study of treeshrew reproduction and behavior with animals kept under conditions where they behaved normally. This was followed a decade later by the only two detailed field studies of treeshrews before my own, both focused on the Malaysian *Tupaia glis*. One of these (Langham 1982) was a capture-mark-release study based on trapping only; the other was based on field observation of a population of marked individuals (Kawamichi and Kawamichi 1979, 1982). Both represented significant advances of knowledge about this extensively studied yet little-known species, and details of their findings will be compared to mine in each chapter below. A short study by Dans (1993) on the Palawan treeshrew in the Philippines reported population estimates, activity, and feeding preferences in captivity.

I was originally drawn to study treeshrews not only because I am fond of small mammals and these were poorly known but also because they were reported by Martin (1968) to have one of the most enigmatic parental care systems known among mammals. The mother *Tupaia* births in a nest remote from her own, and, apart from a visit of a couple of minutes once every other day to suckle her young, she seemingly abandons them. Not only that, she was also reported to lack even the rudiments of the behavioral patterns that ordinary mammal mothers need to care for such helpless altricial infants, such as grooming them or rescuing them from danger (Martin 1968).

This, the ultimate in cursory parenting, was known only from captivity, and there was surprisingly little information about the ecology of wild treeshrews that might help us to understand the natural context of this odd behavior; or, therefore, its function. Intrigued, I went to Borneo to learn how treeshrews of several species lived in the wild, both to see if some feature of their lives might give insight into their "absentee" parenting and to try to describe for the first time in detail the natural history of members of this unique order.

In this study I again asked, What is a treeshrew? but with a different subset of questions. I asked, What is a living treeshrew? How does it act in its environment? What does it do, where does it go, what does it eat, where does it sleep? How does it raise its young? Do different species do these things differently? This book describes what I discovered. I show that "The Treeshrew" does not exist, for each of the six species that I followed has a distinct lifestyle, although all the species share many common features.

I had spent many years in the tropical rainforests of Africa and South America, studying mammals of many kinds. One of my personal goals for this study was to see for myself how the forests of Asia were similar to, or different from, those on the other continents and how communities of mammals resembled, or differed from, each other across the globe. Perhaps from this perspective I could gain insight into how all rainforest communities develop and are maintained.

CHAPTER 2
The Study Species

GENERAL FEATURES OF TREESHREWS

Every organism is captive within the architecture of its inherited physical and chemical structure. Its interactions with the outside world (ecology) are limited by its evolutionary constraints. To establish the physical framework from within which treeshrews must function, in this chapter I review selected aspects of their anatomy and physiology that are especially pertinent to their ecology.

There are two subfamilies of treeshrews, the Ptilocercinae, including only the nocturnal pentail treeshrew (*Ptilocercus lowii*), and the Tupaiinae, including the diurnal genera *Ananthana, Tupaia, Urogale,* and *Dendrogale*. The members of these two subfamilies are anatomically very unlike, with many of their differences linked to their differing activity periods and others perhaps to long separation on independent evolutionary trajectories. My field studies in Borneo included *P. lowii* and five species of *Tupaia*.

MORPHOLOGY

GENERAL CHARACTERISTICS Treeshrews resemble those first, small mammals that skulk in hiding from slavering dinosaurs in the corners of dramatic scenes depicting the Cretaceous. In overall shape they closely mirror some of those ancient mammals: they are of similar size, limb, feet, and tail proportions, and skull outline. The Jurassic pantothere *Crusafontia* (fig. 2.1), reconstructed from its skeleton, has the shape and posture of a living *Tupaia* (Benton 1991). However, the earliest known certain fossil tupaiids are from more than 100 million years later in the Miocene, very late in mammalian evolution, long after most living orders were established. The antecedents of treeshrews have not yet been identified in the fossil record, perhaps because there are few ancient fossils from ancient rainforests of Southeast Asia. The resemblance of treeshrews to *Crusafontia* is testimony, not to any phylogenetic relationships between them, but to the great usefulness of this particular bauplan, or body shape. A treeshrew-sized mammal with five grasping, mobile fingers, a long foot with mobile toes, strong hindquarters, a long tail, and strong humped curvature of the long spine can be a good runner, a good jumper, and a good climber. Its small size allows it to scamper along the branches of trees without any extreme morphological specializations for arboreality (Emmons 1995) yet retain excellent terrestrial running ability. This bauplan evidently evolved early and was so versatile that it is still found in many orders of mammals. The superficial resemblance of treeshrews to squirrels is largely a result of their convergence on a particular body size range, color, and shape that is evidently nearly optimal: squirrels have not changed much since the Oligocene (Emry and Thorington 1982).

The general treeshrew morphology is fine tuned to the individual ecological characteristics of each species in ways that are standard among mammals. Arboreal species have shorter, broader hind feet than do terrestrial ones; the more insectivorous terrestrial species have elongated muzzles, and those that dig have long claws. Pentail treeshrews are particularly adapted for arboreal life: their bauplan resembles that of small marsupials rather than squirrels. Their hands and feet are short, broad, and remarkably strong, with sharp curved claws and digits that can be spread at least 180 degrees across the surface of a branch. On thin supports the hallux is placed above the branch at a wide angle to the other digits, which grip the side of the support (Le Gros Clark 1926; Gould 1978). Their long tail swings in counterbalance when they travel on slender branches (Gould 1978; pers. obs.). Members of the genus *Tupaia* also have structural adaptations for climbing, including first toes that move independently to clasp the top of a branch and ankle joints that can rotate somewhat and allow the treeshrews to descend trees headfirst (Jenkins 1974), as do squirrels. *Ptilocercus lowii* likewise runs down tree trunks headfirst (pers. obs.), and its anatomy shows that it rotates its ankles more

Fig. 2.1. The Jurassic pantothere *Crusafontia*, reconstructed from an almost perfect skeleton (above) (from Benton 1991). A large treeshrew, *Tupaia tana* (below).

fully than does *Tupaia* (Szalay and Drawhorn 1980), but locomotion of living pentails has not been studied in detail.

Pentails seem to prefer to travel connected pathways, without much jumping (pers. obs.), but they seem to be able to jump. On the ground they are apparently quite awkward, and move with "a series of short quadripedal hops" (Gould 1978). *Tupaia* species move with gaits similar to those of most small mammals: at slow speeds they walk, and at a faster pace they gallop or half-bound (Jenkins 1974). They are arboreal leapers and jump to cross small gaps. Even species that forage on the ground are skilled climbers and jumpers.

TEETH   The teeth of treeshrews have been the subject of several studies; below is a summary of some of the published conclusions, principally those of P. M. Butler (1980). Treeshrews have modified tribosphenic molars, of the dilambodont type in Tupaiinae (roughly triangular teeth with three high, sharp main cusps separated by basins). This primitive tooth type arose in the Cretaceous period and is also currently found in didelphid marsupials, but it is not as primitive as the teeth of tenrecs or *Solenodon*. Other insect-eating mammals, such as bats and Insectivora, also retain or have evolved convergent versions of tribosphenic molars. The form of the teeth is more primitive in *Ptilocercus* than in *Tupaia* (Butler 1980). Teeth are evolutionarily plastic, and their form follows function. As in other living taxa, the teeth of treeshrews have a mixture of derived and primitive features.[1]

The long face and muzzle of *Tupaia* species overlies long jaws. The fourth premolar and three molars of treeshrews occlude tightly and form the chewing apparatus. The front portion of the mandible and maxillary forward of the fourth premolar is stretched out such that the anterior premolars and canines are separated by wide gaps (fig. 2.2). The upper

---

1. The concept of primitiveness in features of organisms is relative: of two designs or "states" that a character can have, the most primitive is the one closest to the design that the common ancestor of the two organisms that carry them had. In evolutionary time, the most primitive evolved prior to more advanced (derived) models, and advanced characters were derived through evolution from primitive ones. The scenario is trivially simple, and all of modern evolutionary systematics stands on it; but operationally it can be difficult or impossible to know what design an ancestor had. We are fortunate that mammalian teeth are the best and most often preserved fossil mammal bits, so that for teeth (and little else, I may add) we have an excellent record of much of mammalian history and we can say with some confidence how basic patterns of dentition evolved.

Without it being a law, evolution often seems to favor an increase in a lineage's efficiency in certain tasks, at the expense of others; that is, a change to more "specialized" and less "generalized." A theme in this book will be why I think treeshrews are, or are not, primitive or generalized mammals.

The Study Species

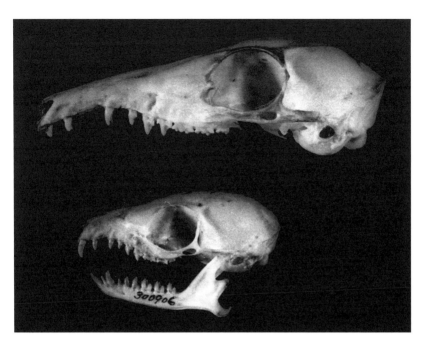

Fig. 2.2. Skulls of treeshrews: large treeshrew, *Tupaia tana* (above); lesser treeshrew, *Tupaia minor* (below).

incisors are large, conical and pointed, curved and caninelike, whereas the lower incisors are broad and procumbent. Together, the incisors form a stabbing and grasping apparatus. The canines and anterior premolars are small, and they do not occlude when the mouth is shut. They act like a pliers adjusted for a wide object, and they can hold or break a large item but cannot crush it finely. The face of *Ptilocercus* is shorter, and the premolars are in lateral contact (but the upper and lower toothrows are still separated by a gap). Perhaps because of the long jaw, the bite at the canines is relatively weak in *Tupaia* (a bite to a human hand scarcely draws blood) but stronger in *Ptilocercus,* which has enlarged temporal muscles as well as a shorter jaw. This arrangement, of stabbing or forcepslike incisors anterior to small, widely spaced canines and premolars that form an open pliers, is found in many insectivorous mammals, such as the true shrews (Soricidae), solenodon (Solenodontidae), elephant shrews (Macroscelididae), and hedgehogs (Erinacidae). There is much variation among these in the forms of the individual teeth. Purely insectivorous rodents have a similar pattern of small stabbing incisors, a dental pattern that is evidently highly effective for handling invertebrate prey.

The gap behind the enlarged incisors may allow the prey to be held firmly while leaving space for the anterior teeth to transpierce it.

In general, tribosphenic teeth have a shearing and puncturing/crushing occlusal function, well suited to reducing the brittle exoskeletons of arthropods, which can be chopped and crushed into particles with a simple vertical chewing motion. But *Tupaia* also has some transverse jaw motion and a broadening of the molars compared to those of early mammals (Butler 1980). These features together may increase the variety of foods that *Tupaia* can efficiently handle (Fish and Mendel 1982). Within the Tupaiidae there is considerable species to species variation in dentition, with the development of stabbing incisors greatest in *Ptilocercus*.

DIGESTIVE TRACT   The parts of a mammalian digestive tract vary in size and structure according to the functional specialization of each part, but generally the more time-consuming and complex the digestion process, the larger and more complex the compartment in which it takes place. The most specialized of all mammalian diets include the structural molecules of the vegetative parts of plants (cellulose). Digestion of these requires an array of microorganisms, a large storage chamber to house them, secretion of a culture medium to keep them viable, digestion time to allow them to work, and area to absorb the products. At the other extreme is a diet of simple carbohydrates (sugars) and proteins (animal tissue), which merely needs a small chamber where enzymes break down ingested tissues into small molecules that can be directly and quickly absorbed.

The digestive tracts of treeshrews are of the simplest gross structure (Hill 1958; see table 2.1 below): the stomach is simple, and the small intestine is narrow and of standard length for a mammal. The caecum is absent in *T. tana* and narrow, smooth, and rudimentary to short in the other species, with that of *T. gracilis* the longest. The large intestine is hardly differentiated from the small (it is difficult to identify in *T. tana*), and it is extremely narrow, smooth, and short.

Among other mammals, only bats as an order possess no caecum, and their whole intestinal tract resembles in its simplicity that of treeshrews, as does that of many insectivorous marsupials. Insectivora likewise have similar simple, tubelike guts. In contrast, the gut of an elephant shrew (*Macrocelides*), which eats some plant matter, has a very large caecum (Mitchell 1916). Mitchell (1905, 1916), with a remarkably modern systematic view, compared the intestinal morphology of an impressive number of mammalian as well as avian taxa: "[The primitive gut] possesses a caecum, or possibly a pair of caeca, homologous with the paired caeca

TABLE 2.1. Mean length (mm) of intestinal segments in treeshrews and some insectivorous and frugivorous mammals (from Emmons 1991). HBL = Head and body length; TGL = total gut length from junction with stomach to anus.

Species (N)	HBL	TGL	Colon	Cecum	TGL/HBL	Colon/TGL	Reference
Treeshrews							
Tupaia tana (5)	210	908	50	0	4.3	0.055	Emmons 1991
T. montana (1)	175	820	30	8	4.7	0.037	Emmons 1991
T. gracilis (1)	133	280	35	11	2.1	0.125	Emmons 1991
T. minor (2)	124	435	42	7	3.5	0.97	Emmons 1991
Fruit Bats							
Pteropus vampyrus	286	1,280	0	0	4.5		Richardson, Stuebing, and Normah 1987
Cynopterus brachyotis	101	656	0	0	4.5		Richardson, Stuebing, and Normah 1987
Artibeus jamaicencis (2)	81	357	0	0	4.4		Emmons 1991
Carollia perspicillata	58	200	0	0	3.4		Klite 1965
Insectivorous Bats							
Molossus molossus	78	123	0	0	1.6		Klite 1965
Pteronotus parnellii	71	186	0	0	2.6		Klite 1965
Squirrel Monkey (omnivore)							
Saimiri sciureus	270	1,307	159	39	4.8	0.12	Fooden 1964

of birds. Unless we accept such a constitution of the primitive or ancestral mammalian gut, we are driven to the much more difficult view that these very definite subdivisions or parts have arisen independently in many groups of mammals. I infer, therefore, that where a mammalian gut-pattern presents less specialization than what I have described as primitive, the condition has come about by secondary reduction" (1916: 188). Mitchell also specifically stated that retained primitive characters were not evidence of taxonomic relationship. Although he did not examine a *Tupaia*, he considered similar morphologies to be secondarily simplified from a slightly more complex ancestor. This view still seems reasonable today. It is noteworthy that the intestinal tracts of several specialized, entirely insectivorous mammals, including anteaters, aardvarks, armadillos, and moles, all have extremely long, narrow, convoluted small intestines and short, narrow, nonsacculated large intestines (Mitchell 1916).

The teeth and digestive tracts of treeshrews are thus similar to those of many other insectivorous mammals and possess a number of both primitive and more derived features.

SENSORY CAPACITIES

Sensory skills are of obvious relevance to ecology: they determine what food can be found, what predators can be detected, and what information can be communicated to conspecifics.

VISION   Treeshrews have large eyes positioned for a wide visual angle but little binocular overlap (see fig. 1.1). The visual system of *Tupaia glis* (the only species whose physiology has been described) is strongly adapted for diurnal activity. The retina is entirely composed of cones (Snyder, Killackey, and Diamond 1969). It has no fovea. The lens is yellow, acting as a yellow filter, a function supplied in primates by a yellow macula (Prince 1956). All-cone retinas and yellow lenses are also found in diurnal squirrels. Yellow pigments in the eye are thought to reduce chromatic aberration and increase contrast (Rodieck 1986). They are found in different tissues (cornea, lens, retinal oil droplets, or macular pigment) of many vertebrates from fish to birds, but among mammals only diurnal squirrels, primates, and *Tupaia* have them (Prince 1956). *Tupaia tana* and *T. minor* have weak, reddish eyeshine (captives we checked at close range), consistent with a lack of tapetum and a day-adapted eye. Behaviorally, treeshrews can discriminate shades of blue, green, and red or yellow but have trouble distinguishing the latter two (Shriver and Noback 1967). They apparently have a dichromatic retina (i.e., containing two visual pigments) (Snyder, Killackey, and Diamond 1969). These features combined (large eye, cone retina, yellow lens, color vision) should give them acute vision by day but poor night vision.

Pentails have large, bulging eyes. Their gross anatomy was described by Le Gros Clark (1926), and little other work seems to have been done on either ocular anatomy or vision in pentails, doubtless because of the rarity of both preserved and living subjects. Clark thought that the retina of *Ptilocercus* was "essentially a rod retina" (p. 1228), based on poorly preserved material. Pentails have far brighter eyeshine than do other mammals, and its pure silver color is unique (pers. obs.). This suggests to me that they have an independently evolved retinal light-reflecting system, evidence of which was apparently noted by Clark: "The fundus oculi shows a certain amount of pigmentation, especially over the periphery and on the nasal side. Elsewhere, the fundus exhibits a beautiful blue-and-silver iridescence in which the fine retinal vessels appear outlined in light blue" (1926: 1227). Their large eyes, probable rod retina, and extraordinarily brilliant eyeshine should give pentails superior light-gathering capabilities. The unique visual system of pentails clearly merits further study.

The behavior of treeshrews in the field suggests that they are strongly visually oriented: as they travel, they scan their surroundings with evident attention, and whereas many mammal species do not seem to notice a motionless person, both *Tupaia* by day and *Ptilocercus* at night usually quickly spot one (which makes them difficult to observe). Their wide visual angle gives them a view above and behind: birds of prey are likely to be their principal predators.

Tupaiids use visual alarm displays. Pentails swing their pendant tails from side to side below the body (Gould 1978) or hold them prominently upward. The broad white tail hairs are highly conspicuous at night. Alarmed *Tupaia* flick the tail, an effective signal at distances as far as the animal can be seen. Treeshrews and tree squirrels have converged in the use of a bushy tail flicked to signal danger, but *Tupaia* flip their tails upward from a horizontal position while most squirrels flick theirs downward from a curled position over the back.

OLFACTION   Like many mammals, including artiodactyla (deer, some bovids) and carnivores (dogs, bears, procyonids), treeshrews have their nostrils set in a wet nosepad that they lick to moisten (fig. 1.1). Also like the former groups, they salivate copiously, and a captive will drool in anticipation of food. Wet nosepads are thought to enhance olfactory sensitivity, enabling canids to find prey by its odor trail and probably helping ruminants to detect both predators and secondary plant compounds in food plants (wet rhinaria seem to be more common in browsers than in grazers). The olfactory lobes of treeshrews are relatively large.

Olfaction is of primary importance for treeshrews, both for foraging (see chap. 5) and for social interactions. They make extensive use of scent marking (Gould 1978; Kawamichi and Kawamichi 1979; Martin 1968), and wild *Tupaia* often sniff along a path, as if following an odor trail.

HEARING   A neurophysiological study of the hearing of *Tupaia glis* compared with a number of primates and prosimians showed *Tupaia* to have an extremely broad frequency range, with maximum sensitivity from 300 Hz to 40 kHz but some response up to 100 kHz, far into the ultrasonic (Peterson, Wruble, and Ponzoli 1968; by recording cochlear microphonics). The treeshrew had hearing about three times as sensitive as that of Old and New World monkeys (mean sensitivity up to fifteen times as sensitive) and somewhat more sensitive and with higher frequency range than that of galagos (*G. senegalensis*).

The ear pinnae of *Tupaia* are short and do not seem to move other than a slight cocking forward during focused attention. Behaviorally, *Tupaia* species are extremely responsive to noise. They flinch dramatically at unexpected sounds that are not loud to humans and focus close attention on noises; but they were not seen to use sound in an obvious way for prey capture. Their small, immobile ear pinnae are not optimal for precise sound localization. However, the auditory sensitivity and high frequency hearing of *Tupaia* (Peterson, Wruble, and Ponzoli 1968) indicate that hearing may be more important for prey capture than we suppose. The recorded calls made by treeshrews are all in the frequency range audible to humans (Gould 1978), but more effort should be made to test for higher frequencies. The reduced pinnae of *Tupaia*, like their bulging, lateral eyes, may give them wide reception to winged danger from above and behind.

Ptilocercus have large, mobile ear pinnae, which swivel together or independently to focus on sounds. In captivity pentails orient toward the sounds of insect prey, including stridulation of Orthoptera (Gould 1978). Large ear pinnae should allow the precise sound localization needed to use this modality for prey capture, and in the wild pentails likely detect some prey by its sounds. Pentails also startle at sounds and quiver and rotate their ear pinnae in response, as do other mammals with acute hearing (bats, desert rodents).

OTHER SENSES   The diurnal treeshrews have virtually no salient facial vibrissae (fig. 1.1), while the nocturnal *P. lowii* has a large spray of them. This may be because pentails not only travel in the dark of night but also climb around inside large tree hollows. *Tupaia* species do neither (although some burrow; see chap. 6), and vision seems to dominate their orientation.

Based on this evidence, the sensory capacities of treeshrews seem to be wider than those of many small mammals, which often do not have all three distance senses well developed (vision, olfaction, hearing). In this treeshrews again resemble squirrels, but elephant shrews and didelphid marsupials also have well-developed vision, olfaction, and hearing, suggesting that the combination of arboreality and diurnality or predatory behavior favors large investment in sensory skills.

THE STUDY ANIMALS

Large islands have dual roles for biodiversity: first, they can provide empty fields for whole new radiations of species, such as Madagascar has for lemurs, Australia for kangaroos, or Hawaii for honeycreepers; and second, the isolation of islands can protect relict forms from extinction by competition or predation from newer species, as Australia has protected the monotremes, or Madagascar the lemurs. Because Borneo only recently became an island again (periodically, when sea levels are low, it is joined by dry land to other parts of the Sunda Shelf), its high diversity of treeshrews, including ten of the nineteen extant species, is more likely attributable to the latter effect than to the former (in situ diversification).

The six treeshrew species that I studied in Sabah (fig. 2.3) cover almost the entire body weight range of the order Scandentia (table 2.2, fig. 2.4), and because they include members of both subfamilies, and both montane and lowland species, they embrace much of the ecological diversity of the order. A thumbnail sketch below introduces each species. The body measurements of these treeshrews show a striking absence of sexual dimorphism (table 2.2). Males and females are virtually identical in size, which is unusual in small mammals or mammals in general.

Fig. 2.3. Portraits of treeshrews: *Ptilocercus lowii* (above), *Tupaia longipes* with radio-collar (below), *Tupaia minor* with radio-collar (above opposite), *Tupaia gracilis* (below opposite).

TABLE 2.2. Treeshrew species studied, standard measurements (mm), means ($\bar{X}$), and standard deviations (SD) of males and females.

Species	Head and Body			Tail			Hind Foot			Ear			Mass (g)		
	$\bar{X}$	SD	N	$\bar{X}$	SD	N	$\bar{X}$	SD	N	$\bar{X}$	SD	N	$\bar{X}$	SD	N
Ptilocercus lowii															
Male	132	5.4	5	180	9.5	5	28	0	5	21	1.1	5	46	5.6	5
Female	134	3.7	6	179	5.7	6	28	0.8	6	21	0.8	6	50	8.7	6
Tupaia minor															
Male	129	6.4	16	159	7.8	15	32	1.5	15	14	1.6	16	52	8.1	16
Female	127	6.3	13	161	9.7	12	32	3.4	12	12	1.1	11	50	9.2	11
Tupaia gracilis															
Male	141	8.9	7	178	4.6	6	39	1.6	7	13	2.9	4	70	12.9	7
Female	142	9.9	5	175	10.7	4	40	1.6	5	12	1.0	3	73	10.0	4
Tupaia montana															
Male	178	16.3	27	152	14.3	26	39	2.2	27	16	2.1	28	127	10.2	26
Female	176	9.8	20	145	9.3	19	39	1.9	20	15	1.6	22	125	10.1	17
Tupaia longipes															
Male	200	16.7	22	194	11	23	50	1.4	23	16	1.9	23	166	21.7	24
Female	199	17.4	31	193	13.9	27	49	1.8	31	16	2.7	30	165	28	26
Tupaia tana															
Male	208	20.5	33	178	10.3	30	46	3.1	33	17	2.2	32	197	22	30
Female	207	10.9	17	176	9.5	17	46	1.9	17	16	2.7	17	210	27	15

NOTE: Data are from specimens in the museum of Universiti Kebangsaan Malaysia, Kampus Sabah, Kota Kinabalu, and the United States National Museum, Washington, D.C. For *P. lowii* and *T. gracilis*, field measurements of individuals captured during the study are also included. All data on *Tupaia* are from Sabah; 10 of 11 *Ptilocercus* specimens are from West Malaysia (Selangor) as no Bornean material was available. Measurements are rounded to the nearest integer.

The Study Species

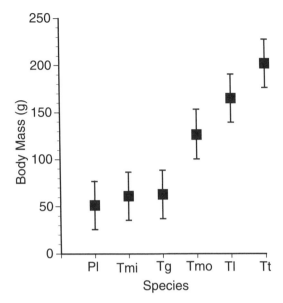

Fig. 2.4. Mean body mass and standard deviations of the treeshrew study species. Pl = *Ptilocercus lowii;* Tmi = *T. minor;* Tg = *T. gracilis;* Tmo = *T. montana;* Tl = *T. longipes;* Tt = *T. tana.*

This implies that body size in treeshrews may have tight ecological constraints and also reflects on social behavior. Among other things, it suggests that wild females finish growth before bearing young, as pregnancy can arrest growth by diverting resources to the young, and that males do not compete physically for females. These ideas are revisited in later chapters.

SUBFAMILY PTILOCERCINAE, PENTAIL TREESHREW, *PTILOCERCUS LOWII* GRAY, 1848

Pentails range across Sumatra, West Malaysia, Thailand, and the lowlands of northern Borneo. This sole member of its subfamily is the only nocturnal treeshrew. It is strikingly unlike other treeshrews in many traits: it has large delicate ear pinnae that can fold back; a prominent, quivering fan of whiskers on the muzzle; soft, gray, woolly fur; and strong grasping fingers and toes. Its most conspicuous feature is a proximally naked tail that is tipped with a large, flat, white flag of distichous hairs. Such a structure is elsewhere found only among gliding mammals.

The stunningly brilliant white eyeshine of *P. lowii* is the brightest I have ever observed in a mammal. Once recognized, this unique eyeshine can readily be identified from a distance of tens of meters, so that these "rare" animals were instead found to be common. Because pentails seldom enter traps, most local naturalists have never seen one.

SUBFAMILY TUPAIINAE, LESSER TREESHREW, *TUPAIA MINOR* GÜNTHER, 1876

The geographic range of lesser treeshrews is similar to that of pentails. This smallest of treeshrews is a uniform olive drab, with a pale shoulder stripe and a relatively slender tail. With its short muzzle and pink nose, it is a cute, Disneyesque version of the more severe-faced larger species. Lesser treeshrews often make a soft peeping sound like little birds, which betrays their location among the foliage. They were very common on both of the study areas and are the easiest species to observe.

SLENDER TREESHREW, *TUPAIA GRACILIS* THOMAS, 1893

Slender treeshrews are endemic to the northern two-thirds of Borneo. They are small, slender, olive drab animals with a pale gray shoulder stripe, a long slender tail, and relatively long legs. This species is almost identical to lesser treeshrews in size and color, and they are often confused both in the wild and in museum collections. Despite their similarity, one can learn to recognize the more drawn out, wraithlike quality of slender treeshrews, even from a distance in the forest. The distinctions between the two species will be emphasized below, because they would seem likely to be in greatest ecological competition. This was the rarest of the *Tupaia* species studied, although it was ubiquitous on my study areas.

PLAIN TREESHREW, *TUPAIA LONGIPES* THOMAS, 1893

Plain treeshrews are endemic to Borneo and found throughout the island. Until recently they were treated as a subspecies of *T. glis,* which is widespread in the Indomalaysian region. This is a medium-sized, dull brown treeshrew with a faint yellow shoulder stripe and a rather slender tail. It was quite common in all areas where we worked, but it is one of the most nervous and wary species, and it is difficult to see for more than fleeting moments.

MOUNTAIN TREESHREW, *TUPAIA MONTANA* THOMAS, 1892

The mountain treeshrew is endemic to the mountains of northern Borneo, where it occurs in mossy montane and cloud forest at elevations above about 900 m. It is medium sized and glossy brown, with a relatively short tail and an almost obsolete shoulder stripe, quite similar in size and color to the plain treeshrew. Its dense, thick coat, protection against cold, wet mountain mists, gives it a rotund shape. It is extremely common where it occurs. We studied it on the slopes of Mount Kinabalu.

LARGE TREESHREW, *TUPAIA TANA* RAFFLES, 1821

The largest of the treeshrews has a geographic range that includes Borneo, Sumatra, and a few small associated islands. It is a handsome animal with a coat of glossy, mahogany red and black, with a yellow shoulder stripe, a black midback stripe on the shoulders, and a tail that varies geographically in color from brown to orange, yellow, or red. It has a long, slender, probing muzzle, elongated foreclaws, and a relatively short, densely haired, bushy tail. Large treeshrews are the most commonly seen and captured species in the lowlands of Borneo.

Recent molecular studies with DNA hybridization and morphometrics indicate that *T. tana* and *T. montana* are sister taxa included within a clade with *T. minor*. *Tupaia gracilis* is a sister group to the latter clade, while *T. longipes* lies outside this combined group on a longer branch (Han, Sheldon, and Stuebing, n.d.).

Although my research was not directed toward discovering the systematic relationships of treeshrews, one of the most intriguing features of this order is its possible value as a model of a primitive mammal, even as a model of what early primates may have been like. As I describe the behavior of treeshrews below, a recurring theme is whether what we see in treeshrews is "primitive," whether their behavior is like or unlike that of other primitive mammals, and whether we can come to any conclusion about what primitiveness is and what its consequences might be.

CHAPTER 3

The Milieu: Field Study Sites and Habitats

One of two Malaysian states on Borneo, Sabah is a quiet, rural region that occupies the island's northern tip, sandwiched between the South China and Sulu seas (fig. 3.1). Borneo was once totally forested, except above the treeline on mountains, but much of its surface is now agricultural, silvicultural, or heavily logged. The sparse human population of Sabah has put relatively little pressure on the land, and the large tracts of forest that remain hold some of the last, best populations of Sumatran rhinos, proboscis monkeys, and orangutans. Beside these well-publicized species lives a rich fauna of little-known smaller mammals (Payne, Francis, and Phillipps 1985). The biota of Borneo reflects its history as part of Sundaland, when during Pleistocene glaciations it was joined by dry land to the continent of Southeast Asia, when sea levels were 160 m lower than at present. Its fauna thus includes species that are the same as, or closely allied with, those of Sumatra, peninsular Malaysia, and Thailand. The biological diversity of Borneo is much enhanced by high mountain ranges with habitat gradients from lowlands to cloud forests, including Mount Kinabalu, the highest mountain in Southeast Asia. About 20 percent of the mammal species of Borneo are endemic to it, including seven of its ten species of treeshrews. All of these treeshrews dwell in the tropical rainforests, and the large geographic extent and elevational variation of Borneo may have provided a field for speciation and persistence of this elsewhere poorly represented family.

The Milieu

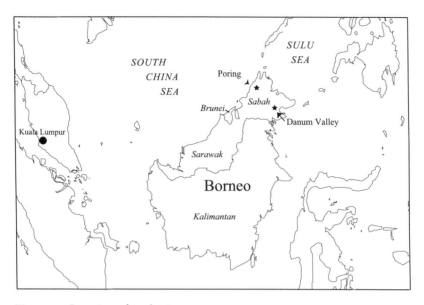

Fig. 3.1. Location of study sites.

Like tropical forests everywhere, the appearance and structure of the forests of Sabah vary considerably from place to place, depending on soil type, slope, drainage, rainfall, exposure, history of disturbance, and biogeographic history. Nevertheless, on well-drained stable ridge tops and slopes grows a forest that is one of the greatest wonders of the biological world. Taller, and with more large trees than other tropical rainforests, the Asian dipterocarp forests are in stature akin to the giant *Sequoia* forests of the United States; but whereas only two or three tree species form the *Sequoia* canopy, more than three hundred species may crowd a single hectare in Borneo (Newbery et al. 1992; Primack and Hall 1992). Many emergents (trees that emerge above the canopy) reach 50 m in height; a few tower to above 75 m (fig. 3.2). These giants are mostly of the dipterocarp family (Dipterocarpacae) and a legume (*Koompassia excelsa*). However, where soil conditions are unfavorable to the dipterocarp giants, a Bornean forest looks much like tropical rainforests on other continents worldwide: it may be tall and open below; irregular in height and undergrowth; or short and sparse in canopy and snarled with festoons of lianas or thickets of vicious rattan climbers. Treeshrews live in all forest types on Borneo.

Fig 3.2. An immense, canopy-size dipterocarp tree on a logging road near Poring. Note Alim Biun and Hajinin Hussain standing in the road to the left of the base of the tree.

PORING HOTSPRINGS RANGER STATION

The first phase of field studies was from March to August 1989, at Poring Hotsprings Ranger Station (Poring), in Kinabalu Park, Sabah, Malaysia (06°03′N; 116°42′E; fig. 3.1), under the hospitality of the Sabah Parks Department. Two of the park staff, Alim Biun and Hajinin Hussain, collaborated full time in the research during this field season, to the great benefit of the study. Alim and Hajinin cheerfully endured barbaric work schedules (sometimes radio-tracking from 4:30 in the morning to 8:30 at night) and taught me much about the animals and forest. At Poring we worked on two study plots, a "low" plot on a ridge crest at the base of the Eastern Ridge of the Mount Kinabalu massif and a "high" plot on a ridge 2.5 km northwest from Poring, above the trail to Langanan Falls.

The low plot occupies a small side ridge separated by a steep ravine and stream from the main Eastern Ridge. We chose this site because it included the largest relatively flat area of lowland forest in that section of the park. All vegetation below it was secondary or agricultural. The plot began at about 535 m elevation at the lower limit of the mature forest and extended 500 m in a wedge to where the ridge narrowed to about 50 m at 720 m elevation. We worked along a trail above this 300 m farther, to 800 m elevation. On this area we opened and used a small trail system for radio-tracking, which was flanked by steep ravines on each side, and we trapped treeshrews in a grid on the wider, lower part (fig. 3.3). The flattish triangle within our trails measured 6 ha and was 200 m wide at the base and 460 m long. We followed treeshrews in this plot throughout our study at Poring.

This lowland plot had vegetation characterized by a high canopy of emergents, dominated by tall dipterocarps on the higher well-drained ridge tops and edges, with generally smaller trees such as oaks (*Lithocarpus* spp., *Castanopsis* spp.) occupying the lower portions. The understory ranged from bare and open to vine-choked ravines and thickets of saplings in old lightgaps (openings caused by treefalls). Most of the study plot had a fairly open understory. A high diversity of dense rattan palms scrambled throughout much of the forest, and groves of understory *Licuala* sp. and other palms grew at 700 m on slopes (fig. 3.4). Where light entered on ravine sides or rocky slopes, curtains of vines draped the steep rock faces. Except in dank ravine bottoms, there were scarcely any trunk or branch epiphytes on the trees at this elevation, apart from isolated baskets of giant staghorn ferns *(Platycerium* sp.). The forest floor had a good cover of leaf litter, except where this was swept away

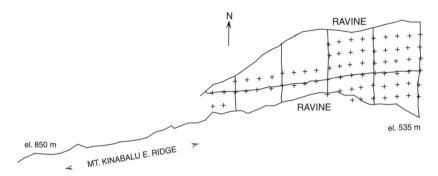

Fig. 3.3. Map of trails and trap sites (crosses) on the lower plot at Poring.

by water runoff. The soil here at the mountain base was of brown to red clay, slippery as grease when soaked, interspersed with granite rocks. The only surface water was in streams at the bottom of ravines at either side and in a few wallows made and maintained by bearded pigs (*Sus barbatus*). When they held water, these wallows were water holes used by mammals, birds, and amphibians.

The Langanan (high) plot was likewise chosen for its flattish profile and workable vegetation, after initial trapping showed it to include many montane treeshrews. It also was on a ridge top, in the shape of a rough equilateral triangle encompassing 3.8 ha. It ranged in elevation from 930 to 1,000 m. We filled the area with lines of traps spaced at 25 m intervals. We worked in this plot only to study mountain treeshrews (*T. montana*), from 29 June to 31 July 1989. To reach the high plot to start work before dawn, we had to leave our house at 0430 h and walk for an hour.

The forest on the high plot is a closed canopy of small- to medium-diameter trees, generally small-leafed, growing on a rocky substrate (fig. 3.5). The understory is dense and characterized by numerous, prehensile rattans that grab one's skin and clothing and enforce cautious movement. At this elevation condensation from the afternoon mists, which daily shroud Mount Kinabalu during many months of the year, is evident in an abruptly increased density of trunk and branch epiphytes such as moss, ferns, and orchids; moss on the rocks and tree trunks; and the formation of a spongy mat of tree roots and moss lying above the soil surface. At the level of the plot this springy surface root mat was just beginning to develop and quite thin. At higher elevations it can become several feet thick, with the moss-covered root mat entirely above ground

The Milieu

Fig. 3.4. The forest understory at 600 m on the study plot at Poring.

and extremely difficult to traverse (one repeatedly drops hip-deep into unseen holes full of black muck), a reason we chose to work at the low edge of this altitudinal transition.

The lands skirting Mount Kinabalu are quite densely occupied by Dusun people, for whom hunting is an important tradition. Poaching in the park was intense and difficult to control, and large game mammals were much reduced or in some cases extirpated in the vicinity of Poring, so that the fauna there appeared to be diminished at the time of our study. Much of the forest outside of the park has been cut, leaving few natural areas for legal hunting, and there were no significant tracts of lowland forest workable from Poring.

DANUM VALLEY

In 1990 I moved the project to Danum Valley Conservation Area (Danum), in eastern Sabah near Lahad Datu (4°58'N, 117°48'E; fig. 3.1), chiefly because the steep, ravined terrain at Poring made it very difficult to track the weak transmitters carried by small mammals. It also seemed preferable to work in a site with less hunting and a more intact and tranquil fauna. I studied treeshrews in the field at Danum Valley Field Centre from 29 August 1990 to 2 October 1991. At Danum an able botanist,

Fig. 3.5. The forest at about 900 m on the high study plot at Poring, habitat of *T. montana*.

The Milieu   31

Fig. 3.6. Helicopter view of Danum Valley Field Centre. The treeshrew study plot is in the forest above the river in the right half of the photo. The slender suspension bridge can just be seen.

Elaine Campbell Gasis, collaborated with me throughout the year to collect and identify monthly fruit samples. Mohammad Dzuhari, from World Wildlife Fund Malaysia, interned as my assistant from December to March and kept the research going from 3 January to 3 March 1991, when I was absent from the field for an expedition to Ecuador. The Field Centre at Danum is one of the most luxurious and well appointed in the world, with fine new buildings surrounded by the rainforest (fig. 3.6).

Danum Valley Field Centre and its 43,800 ha reserve of pristine forest, which comprises the Danum Valley Conservation Area, has been the base for extensive field research. This has allowed my studies on treeshrews to be embedded into a far more complete ecological background than could have been possible elsewhere in Sabah. Some of these studies are collected in a recent publication (Marshall and Swaine 1992) that includes outlines of the history of the field station and its forests, soils, and climate (Marsh and Greer 1992). Although poaching has severely

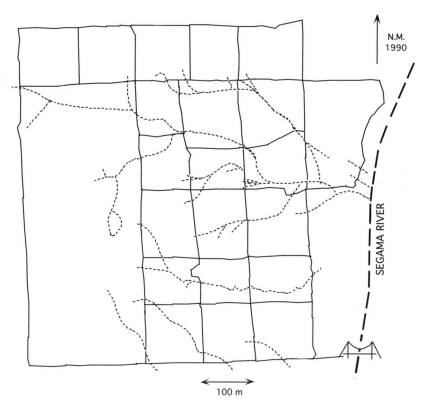

Fig. 3.7. Map of the trails (solid lines) and streams (dotted lines) on the study plot at Danum Valley.

reduced Sumatran rhino numbers within the conservation area at Danum, the remainder of the vertebrate fauna was intact and little disturbed by humans. The ecological community that I studied was in as "natural" a state as could be found anywhere in Asia.

The study plot at Danum is within the conservation area adjacent to the Segama River, just north of the bridge, and ranges from about 450 m elevation at the river to 650 m elevation at its top (northwest) corner. It occupies the east-facing slope of the west (left) bank of the river basin and includes two permanent and several temporary small tributary streams. The terrain is a rolling series of erosion gullies and ridges perpendicular to the river on the flank of a long, gentle ridge that reaches 900 m elevation about 2 km west of the river. I chose this as the focal study area because it had already been the base plot for studies of the bird fauna (Lambert 1992) and litterfall and litter fauna (Burghouts et al. 1992) and was adjacent to a botanical tree plot (Newbery et al. 1992). It was also partially subdivided by trails at 100 m intervals and conveniently close to the Field Centre, so that it was a quick walk (10–20 min) to reach any part of the plot to start radio-tracking before dawn (fig. 3.7). Within the main trails used for radio-tracking, the study plot had an area of 17.74 ha, but the treeshrews I followed used a cumulative area of about 37 ha around these, reaching to at most 600 m west of the river and about 1,000 m north of the bridge.

The vegetation of the study area is classified as lowland evergreen dipterocarp forest (Marsh and Greer 1992). It has a high tree species diversity, with 511 species (>10 cm DBH = 2,248 individuals) identified in a 4 ha parcel next to my study plot, of which the 2 ha halves had 388 and 387 species, respectively (Newbery et al. 1992). Dipterocarpaceae was the second best represented tree family in the parcel, with 9.3 percent of individuals (about 100/ha); but Euphorbiaceae strongly dominated, with 27.5 percent of all trees (about 300/ha), and it would more appropriately be called a euphorb than a dipterocarp forest. My adjacent study plot has large disturbed areas and also seemed to have fewer large trees than the Newbery parcels, perhaps because of the effects of wind and erosion (treefalls) on the river-facing slopes and because the Newbery plots are on a quite flat, well-drained ridge top.

The treeshrew study area is a mosaic of tall open forests on well-drained ridgelines, intersected with moist, densely vegetated gullies and a small ravine. The canopy generally has a high layer of emergents (40–60 m tall) and large trees (30–40 m) that are separated by wide gaps through which light penetrates to fall on a closed subcanopy formed by the tops of small trees and lianas many meters below (fig. 3.8). The true, connected closed canopy is often very low, at 6 to 10 m, although there is large variation in its height. This structure, with lack of closure of the high canopy, may be created or favored by the hilly terrain, which precludes formation of a uniform canopy surface and allows some side penetrance of light. The understory on the plot includes almost no rattans or other palms and varies between open on well-drained ridge tops to totally closed in old treefall gaps. There are few epiphytes, and the forest floor has in most places a well-developed standing layer of leaf litter (production about 3.4 t/ha/yr; Burghouts et al. 1992). The open river and its valley exposes the basin to winds that are desiccating during dry months. Four large areas of about a third to half a hectare each have no tree canopy and are smothered with nearly impenetrable tangles of old lianas. The largest such area covers a rocky substrate much like a boul-

Fig. 3.8. Helicopter view of the forest canopy on the study plot at Danum Valley. Note how tall, isolated trees emerge above a lower, more closed tree canopy below.

der field and may be edaphic, but others seemed to be self-propagating treefall zones, where the gap from a huge treefall causes other trees to die from wind and exposure and increases gap size, while exuberant vine growth suppresses tree regeneration.

For comparison with results from mature forest at Danum, I made a brief trapping survey in logged forest outside the conservation area 1 km northeast of the field center. This forest had been selectively logged from primary forest in 1989. It was a mosaic of relatively little damaged areas that still had some tree canopy and extremely disturbed sections, open to the sky, with many vine blankets and stands of *Macaranga* and *Vismia*.

RAINFALL AND FRUITING PHENOLOGY

There were no climate or rainfall records for Poring Hotsprings Ranger Station at the time of my study, but detailed climate records have been kept since 1985 by the stream hydrology project at Danum Valley (see Marsh and Greer 1992). Mean annual rainfall there from 1986 to 1990 was 2,822 mm. Monthly rainfall shows no strongly predictable pattern,

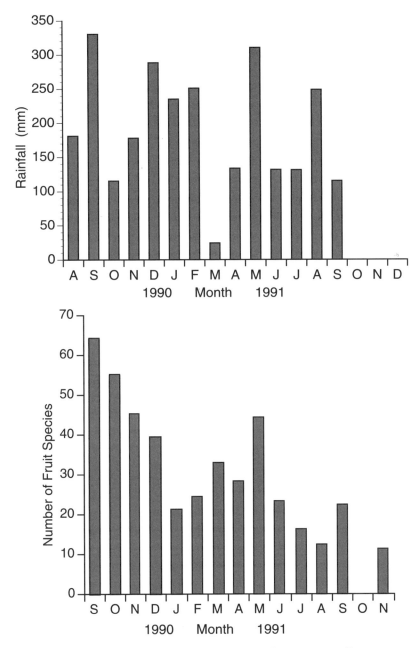

Fig. 3.9. Monthly rainfall during the study period at Danum Valley (top) (courtesy of the Danum Hydrology Project); number of fruit species collected per month on study plot trails (bottom).

but April is usually the driest month, with May and July–September also often dry; November–January and June tend to be wettest. The monthly rainfall during the study year at Danum Valley (fig. 3.9) was atypical in the context of previous years in that there was a drought in March 1991.

To evaluate the seasonal patterns in a treeshrew resource at Danum Valley, Gasis worked with me to collect and identify all fallen fruit from a transect that included the grid of 4.5 km of trails encompassing most of the home ranges of the radio-collared study animals. The complete list of fruit species that we identified is compiled in Appendix I.

The overall fruiting phenology showed two peaks of fruit abundance during the calendar year (fig. 3.9), centered in September and March, the rainiest months during the study. By good fortune, the start of our work at Danum coincided with the first major fruit-masting episode recorded at the Field Centre since its inception in 1984, so we were able to document the consequences of a rare fruiting event. Unfortunately we missed the beginning of the masting phenomenon and only monitored its decline from the peak period (September–December 1990; fig. 3.9). More fruit species were collected in each of the four months during the masting period than in any of the ten following months. The number of species of fallen fruit recorded during masting in September 1990 (63) was three times that of September 1991 (22), and November 1990 had four times as many fruit species as the same month in 1991.

APPENDIX: METHODS

Most methods germane to specific topics are detailed in appendixes in the relevant chapters, and only a few general techniques are given here. The statistical analyses and some graphs were done with Systat© 5.2 for the Macintosh, and most graphs were drawn with DeltaGraph© Pro 3.

Fruit Sampling Methods. We collected fallen fruit during two days near the beginning of each month from September 1990 through November 1991, except October 1991. All fruit in each 100 m trail section was collected together into a separate plastic bag, so that information on the number of fruit species, and total weight of fruit, could be evaluated in terms of distance. Our fruit survey is but a subsample of what was available on the study area. Although there are a number of obvious biases created by the methodology itself, the method should give a good reflection of relative monthly values (fig. 3.9B).

The outstanding feature of the rainfall records for the year before the masting phenomenon was an exceptionally wet May and June, with 600 to 700 mm of rain falling in each, almost double the usual amount, following a very dry April (about 30 mm). These rains perhaps triggered the mass flowering that subsequently produced massive fruiting.

The Milieu

In August and September 1991 dense, acrid smoke, evidently from burning rice fields in Kalimantan hundreds of kilometers away, blanketed Danum and most of Sabah. Airports were temporarily closed, and visibility was reduced to a couple of hundred meters. It is possible that blockage of the light could have influenced plant phenology, or rains falling through smoke may have acidified the streams, but no data are available.

CHAPTER 4

Treeshrews in Their Habitat

The first task at the start of any field study is to find the study species. This can be the most difficult part of fieldwork, and projects sometimes founder at this initial point. Among the most satisfying moments to a field biologist is the instant of realization that one can predict where an animal (or plant) can be found. At that moment one has begun to understand the organism. This account of the results of fieldwork on treeshrews starts at its logical beginning and first field question: where are the treeshrews? I answered this simply, by learning to recognize the species by sight in the field and observing where they were to be seen as I walked trails throughout the study areas (see the chapter appendix for more on methodology). It took several months of trapping and watching before I was fairly confident of field identifications, but accuracy and speed continued to improve throughout my stay in Sabah.

The paths that treeshrews travel in the forest outline where they seek and find food. Their physical distribution thus describes ecological features with a potential to define the roles of species in the environment. This is true of their use of the habitat both on a small, detailed scale, such as the exact height they can be found on the side of a tree, and on a larger scale, such as where they install their home ranges among all the possible habitat types. This chapter thus presents information about both height and habitat distribution of the study animals.

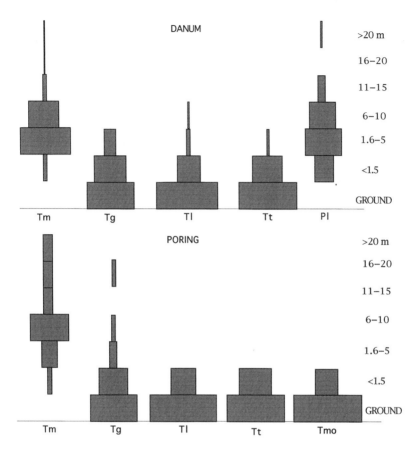

Fig. 4.1. Height distributions of sightings of treeshrews in the two study areas, in percent of observations. First four species are arranged in increasing order of mass. Tm = *T. minor*; Tg = *T. gracilis*; Tl = *T. longipes*; Tt = *T. tana*; Tmo = *T. montana*; Pl = *Ptilocercus lowii*.

VERTICAL USE OF THE FOREST

Field sightings immediately showed that the treeshrew species differ in how they make use of the vertical dimension of the forest. Only the two smallest species, *T. minor* and *P. lowii*, forage chiefly arboreally (fig. 4.1). The four larger species are terrestrial, and in from 75 to 90 percent of all sightings, they were below 1.5 m height in the forest. This low stratum from the ground to 1.5 m encompasses the complex fretwork of fallen logs and branches that are used as refuges, pathways, and foraging sites by small terrestrial mammals.

The minor height differences among the four terrestrial *Tupaia* species may seem trivial when tabulated (fig. 4.1), but I believe that they indicate true differences in fine details of the way that each species forages. In Borneo most treeshrew species are difficult to observe in the field, and the numbers of identified sightings were in some cases too few to support statistical comparison. However, during weeks of following individuals of each species by radio-tracking, all glimpses of the animals and all inferences that could be made about the radio signal heights confirmed the results of direct sightings. A field note sketch of each species in the forest shows how it uses the vegetation column and substrate.

PTILOCERCUS LOWII

10/20/90; 1950 h: A *Ptilocercus* is running rapidly up and down from 0–8 m on rattan stems that have many dead leaves, in dense growth. 10/27/90; 1940 h: I watch a *Ptilocercus* for about 5 min as it runs up and down a vine-draped tree from the canopy to 2 m. It went down, then up, then down, moving rapidly on the main trunk, never still.

I saw pentail treeshrews forty-eight times, always above the ground (excluding observations on the nest tree; see chap. 6). Pentails at Danum seemed to forage across the whole vegetation column in a quite even distribution from top to bottom. It was clear from radio-tracking that *P. lowii* spent more time than is shown in figure 4.1 in the higher canopy layers, above the height where they could be detected at night by observation. Pentails were also sensitive to the light of a headlight and avoided it, which may have inhibited some animals from descending when I was below. I sometimes watched them with red-filtered light, but it was difficult to use this while radio-tracking for many hours.

TUPAIA MINOR

5/13/89; 0700 h: *T. minor* F112 seen foraging in vines among dense, high canopy foliage at about 25–30 m. She stays out on branch tips foraging among and shaking leaves, moving quite slowly from branch to branch. Lost from view 0710 h. 0721 h still in same tree, 0725 h she chin rubs on branch, goes through foliage to a neighboring tree, then back to first tree. 0735 h still foraging very actively, vigorously shaking foliage, then comes down to neighboring tree 10 m above me, foraging in living and dead leaves, mostly on outer canopy layer. I've seen her eat several items, but not what. Out of sight 0740 h.

Tupaia minor was by far the most commonly seen species, with about two hundred recorded sightings during the study. Radio-tracking confirmed that lesser treeshrews spent most of their foraging time in the densely connected and densely leafed subcanopy, but they sometimes spent hours climbing high out of sight in vine-enveloped emergent trees. I never saw a lesser treeshrew on the ground, although they sometimes ran along low fallen logs, especially to cross treefall gaps. This species can sometimes be caught in traps placed on the ground, but I believe that they are drawn down to them by the smell of bait. Most purported sightings of this species on the ground in Borneo are likely to be cases of mistaken identity and to be slender treeshrews instead. Arboreal behavior explains the large number of observations: a much greater area of aboveground than ground-level space is visible to a human observer, so arboreal mammals are far more easily seen than terrestrial ones. Moreover, safe in a tree, lesser treeshrews are the least wary of any of the study species and fearlessly continue their activities in the presence of large, intently staring, terrestrial bipeds. This was the only species that I was ever able to follow visually for more than about 10 minutes. The relative use by *T. minor* of vegetation levels above 10 to 15 m, where animals may disappear from view, cannot be inferred from figure 4.1 (see methods).

TUPAIA GRACILIS

> 5/26/89; 1317 h: A *T. gracilis* comes right by me, walking on the ground foraging, nosing the ground and low foliage. It jumped onto a log and walked nose to nose into a large skink: each mildly backed off a couple of inches and went their ways, the *gracilis* sniffed a dead leaf hanging from a shrub. 5/30/89; 1338 h: A *gracilis* comes foraging under a log, sniffing both ground and overhead leaves, mainly surface foraging in a highly zigzag path. 4/26/91; 0634 h: A large unmarked *gracilis* traveling on the ground runs 2 m up a sapling, then forages up and down a vertical sapling to 3 m, then down to the ground.

Slender treeshrew sightings were recorded seventy-one times, three-fourths of them at 1.5 m or below. During more than three hundred hours of radio-tracking, I noted individuals apparently climbing into the canopy only three times, each for only a few minutes. This species is basically a terrestrial forager that always travels at about ground level but does some exploring in the lower levels of undergrowth vegetation, as described in the field notes above. Slender treeshrews like to travel along the tops of fallen logs, branches, and roots, even small ones close to the ground. Because treeshrews are much more visible if they climb just 1 to 2 m off the ground, animals off the ground are likely to be overrepresented relative to those on the ground in this and similar data sets. There is no way to quantify this bias, but it is certain that it leads to underestimating animals on the ground relative to those 1 to 5 m above it. The strong terrestriality of most treeshrews may thus be even more pronounced than indicated in figure 4.1.

TUPAIA MONTANA

12/7/89; 1742 h: *T. montana* M151 crosses the trail nosing among leaf litter, jumps onto a log and runs along it, then jumps down and forages along the ground. 24/7/89; 0750 h: *T. montana* M143 comes up to a trap, sniffs around it, goes inside a hollow log, then goes away walking on ground.

Mountain treeshrews were studied for only a few weeks, when merely eight sightings were recorded, all below 1.5 m. These treeshrews seem to be entirely terrestrial foragers. They are extremely numerous and are often heard calling, but because they are at ground level in a dense and cluttered habitat, they are difficult to spot, except on roads or cleared trails. Unlike some other *Tupaia*, mountain treeshrews had little tendency to jump up onto higher objects to view their surroundings when disturbed. The presence of a look-alike squirrel (*Dremomys everetti*) in its habitat gives sightings a greater uncertainty.

TUPAIA LONGIPES

6/21/91; 0725 h: A *T. longipes* with radio comes foraging along the ground surface, jumps up on a log and runs down it, picking up something in its mouth and eating it without stopping. 11/7/90; 1341 h: *T. longipes* foraging rapidly along the ground nibbling at things on litter surface. 7/13/91; 1002 h: *T. longipes* F56 foraging in dense vines, repeatedly climbing up and down on them from 0.5 to1.0 m.

The sixty-eight recorded observations of plain treeshrews show them to be intensely terrestrial. Only three individuals were seen above 1.5 m, but, significantly, all three were foraging for arthropods, once up to 8 m. Thus, although it is rare, *T. longipes* forages to a small degree on living vegetation, especially trunks and vine tangles just above the ground. This species was the most cryptic, nervous, and wary of all those studied, and it was rarely possible to watch them for longer than seconds, except from blinds at artificial bait sites.

TUPAIA TANA

12/11/90; 0725 h: *T. tana* F54 comes foraging around the viny buttress roots of a tree, goes into a dark cavity under an upended root pile, stays 2 to 3 min, then comes out and crosses the trail on a log, goes up into a log pile, goes down into another dark cavity under a log and disappears. 0820 h: She comes foraging down a steep bank by large buttress roots in a stream, nosing very slowly along sides of roots, and under leaves, quiet and inconspicuous.

Large treeshrews were the second most often seen species, with 104 records of height use. They foraged on the ground and on or under logs and piles of dead brush, but I never saw them using living, aboveground vegetation as a foraging site. However, they often jumped up on saplings or trunks to scan the surroundings when they detected a disturbance, and they also often traveled on slightly elevated pathways along branches and logs or jumped between vertical saplings (perhaps to keep the observer in sight). These treeshrews were calmer than the other terrestrial forms, and some individuals became so tame during radio-tracking that they sometimes let me watch them for a few minutes.

SUBSTRATE USE

To define habitat use more narrowly, I recorded the substrate on which a treeshrew was at first view (table 4.1). These data give an idea of general tendencies, but because the categories recorded were crude and somewhat arbitrary, and I did not always note substrate, it lacks statistical power. The results show that (1) *T. longipes* and *T. tana* use virtually identical substrate arrays, but *T. gracilis* uses more living vegetation than the other two large terrestrial species while *T. montana* may use less (these trends mirror the height distributions described above); (2) the "sibling" species[1] *T. minor* and *T. gracilis* are as strikingly different in substrate use as they are in height distribution; and (3) the two, similar-sized arboreal species, *P. lowii* and *T. minor*, use the vegetation volume in quite different ways: whereas lesser treeshrews favor vines, pentails use large vertical trunks and the tops of understory treelets. *T. minor* is notable for its intensive use of viny habitats. These include all types of vine-covered vegetation, such as horizontal blankets of pure vines over old treefalls, vertical ones down banks and edges, a few strands connecting midstory tree crowns, or deep layers wrapping emergent trunks.

These distinctions in substrate use between *T. minor* and *P. lowii* are also linked to a difference between them in movement pattern (table 4.2). Lesser treeshrews foraged about two-thirds of the time traveling in a ba-

---

1. Species so similar externally that they are difficult to distinguish by eye.

TABLE 4.1. Distribution of sightings of treeshrews on various substrates.

Species	Ground	Log	Branch	Liana	Trunk	Treelet	Shrub
P. lowii	0	1	2	10	19	15	
T. minor	0	4	4	148	5	6	3
T. gracilis	37	11	5	14		2	1
T. montana	6	2					
T. longipes	47	14	1	7	1		
T. tana	46	16	1	4			

NOTE: Because several substrates were often sequentially used during an observation, and choice of which to score was arbitrary, data only indicate trends. Log category includes treefall brush piles; liana includes vine-covered trees. Trunk includes vertical but not fallen tree.

TABLE 4.2. Chief direction of movement of treeshrews observed foraging arboreally. Percentages of sightings (N).

Species	N	% Horizontal	% Vertical
P. lowii	35	11	89
T. minor	75	63	37

sically horizontal direction and the other third climbing vertically up and down, often in hanging curtains of vines. Pentails, in contrast, spent almost 90 percent of their time running vertically up and down, often on large, bare trunks.

PREVIOUS STUDIES OF FORAGING LEVELS

The basic foraging level of most treeshrew species, either arboreal or terrestrial, was noted from time to time by naturalists, starting with early collectors: Raffles (1822) and Banks (1931) described *T. tana* as terrestrial; Ridley (1895) stated *T. glis ferruginea* was terrestrial; and Banks (1931) noted that *T. montana* was terrestrial. Davis (1962) published the most comprehensive collector's account, which described many aspects of natural history. He recorded the terrestrial habits of *T. tana, T. longipes,* and probably *T. dorsalis;* and arboreal habits and preference for "small trees and vines of the lower middle story" of *T. minor* (p. 49). *Ptilocercus* was recorded as arboreal by Harrison (1962). Davis (1962) believed *Tupaia gracilis* was similar to *T. minor* in height preference, as did Payne, Francis, and Phillipps (1985). Only for slender treeshrews, therefore, are my foraging level results entirely new.

Kawamichi and Kawamichi (1979) studied common treeshrews (*T. glis*) by observing a dense population of marked individuals on the island of Singapore. Their vast number of observations (N = 1,905) recorded *T. glis* below 1.5 m in 95.6 percent of sightings, with the balance consisting mostly of short bouts of insect foraging in low vegetation. This closely matches the height distribution and behavior of *T. longipes* (thought by some to be but a subspecies of *T. glis*).

In a short field study of *T. minor* in West Malaysia, D'Souza (1972) found a lower height distribution, with 11 percent of the animals on the ground, than that I report here (none on the ground). Since his few sightings (N = 35) were all made from three fixed observation points, the sample may not be broadly representative of continental populations. However, perhaps *T. minor* uses the ground to a greater extent on the mainland, where *T. gracilis* is absent, than on Borneo, where it is present.

EVOLUTIONARY NOTES

The capacity for arboreal leaping (without gliding) is developed in few of the living mammalian orders: only the treeshrews, primates, carnivores, and one family of rodents (squirrels) have evolved this talent to a significant degree (Emmons 1995). In the latter three taxa, arboreal leaping is associated with speciose radiations of canopy-living mammals (primates, squirrels, and martens, genets, palm civets, and procyonids). All treeshrews are adept climbers and leapers, but unlike virtually all other families of arboreal-leaping mammals, truly arboreal lifestyles are rare among them. Perhaps only two or three of the eighteen species of *Tupaia* are arboreal, yet all are highly skilled climbers. When disturbed, *T. tana*, one of the most terrestrial species, even travels with a kind of arboreal cling-and-leaping, jumping with ease from one understory sapling to the next, perhaps to keep the perceived threat in view. Nonetheless, as a group the treeshrews are a terrestrially foraging order, with arboreality being the exceptional lifestyle.

This raises an old and interesting question (see Jenkins 1974): Is arboreality therefore the ancestral condition of Scandentia and terrestriality more derived? Is the leaping skill of terrestrial treeshrews the reflection of a more arboreal history? Because *Ptilocercus lowii*, the most morphologically primitive treeshrew species (Luckett 1980), is strictly arboreal, and all species retain morphological adaptations for arboreality, including the complex specific functional adaptation of rotating ankle joints (Szalay and Drawhorn 1980), it seems almost a certainty that the living species all derive from an arboreal ancestor. Why, then, are most species terrestrial? In the next chapter we will see that what treeshrews eat has a bearing on the answer. In a pattern contrary to that of primates, treeshrews have apparently come down from the trees.

HABITAT RANGES

Months of work in small study plots that are only a few home ranges wide gave but a limited, snapshot view of the habitat preferences of treeshrews, and I wanted a broader, more panoramic vista. On spare mornings I walked many kilometers of trails through surrounding forest, keeping records of treeshrews and all other mammals sighted, but the sighting frequency for treeshrews was so discouragingly low in relation to the numbers I knew were there that it was obvious that direct observations could supplement, but not replace, captures, and I could evaluate treeshrew distributions only by trapping. There was time for only short trapping projects, one at each study site, to investigate two questions: how do the montane and lowland species interdigitate on the elevational gradient, and does selective logging influence species composition? The other mammal species recorded in these habitats are listed in Appendix III.

THE ELEVATIONAL GRADIENT
OF TREESHREWS ON MOUNT KINABALU

How does *Tupaia montana* replace its terrestrial congeners, *T. tana, T. longipes,* and *T. gracilis,* on an altitudinal transect? Do the three lowland species drop out together at the same elevation, or separately, at different ones? Does *T. montana* overlap the others, or are the species entirely separated in space? Is there a vegetational change at the elevational limits of treeshrew species? These were questions I hoped to answer.

The slope on our low study plot at Poring was so steep that it reached 850 m elevation only 750 m in a straight line above its base trail at 535 m elevation (920 m by trail over the ground). The uppermost trap of our regular trapline was at 725 m elevation. The home ranges of radio-tagged individuals captured on the trapline reached 850 m for *T. longipes* F133; 770 m for *T. tana* F106; and 780 m for *T. gracilis* F105. On the high plot at Langanan ridge, we ran a single line of thirty traps for three days (90

TABLE 4.3. Elevational limits observed in *Tupaia* species on Mount Kinabalu, near Poring, and the number of individuals known to inhabit the Langanan (high) study plot at about 1,000 m; the pair of *T. minor* as seen, the others as captured.

Species	Highest Elevation (m)	Lowest Elevation (m)	Individuals on High Grid
T. minor	970		2
T. gracilis	780		0?
T. longipes	970		1
T. tana	1,000		1
T. montana		890	11

trap-days), from 880 m elevation (close to the upper limit reached by treeshrews on the low plot) to 1,000 m. Following that, we placed the grid of sixty-nine traps from 950 m to 1,000 m elevation (466 trap-days).

We found that at 900 to 1,000 m on the Langanan plot we were just at the upper limit for lowland *Tupaia* species and at the lower limit for *T. montana* (table 4.3). At this place the highest home ranges of *T. tana* and *T. longipes* overlapped the lowest ranges of *T. montana*. We did not see any of the montane, smooth-tailed treeshrews (*Dendrogale melanura*) that occur at 1,200 m elevation at Kinabalu Park Headquarters.

The highest *T. tana* captured at Langanan was a subadult female who ranged from 905 m to 1,000 m. Because the elevational gradient is so steep on Kinabalu and home ranges of the terrestrial treeshrews are so large (300–600 m long; see chap. 8), the addition or subtraction of a single home range at the altitudinal limit could shift it 100 to 300 m in elevation. One would therefore not expect a very precise elevational cutoff if ranges are determined by habitat transitions. *Tupaia gracilis* was never identified above the top of F105's territory on the low plot (780 m), but on the Langanan plot I once saw an unidentified tiny treeshrew (*T. minor* or *gracilis*) running repeatedly up and down a tree from the ground at 995 m, presumably building a nest in the pouring rain. A pair of lesser treeshrews lived in the middle of the Langanan plot, and there could have been others higher, but I saw none during walks at higher elevations.

DISCUSSION OF ELEVATIONAL DISTRIBUTION

The distribution of altitudinal records for treeshrews in Borneo summarized by Payne, Francis, and Phillipps (1985) is similar to that which

I found at Poring, with the expected variation arising from different local climate patterns on different mountains. The lowest territories of *T. montana* overlapped the territories of *T. longipes* and *T. tana,* so there did not appear to be direct exclusion (interspecific territoriality) by *T. montana* of the other two large terrestrial species at the transition zone. The lower elevational limit of montane treeshrews coincides with the elevation at which it becomes moist enough for the formation of a ground surface mat of rootlets and moss and a generally increased epiphyte load. On many tropical mountains this occurs somewhere between 900 and 1,500 m elevation, where the air becomes just cool enough to cause frequent condensation in mists and clouds. This transition has an abrupt change in vegetation structure, accompanied by a turnover of small mammal species, in the Andes (Cadle and Patton 1988), the Philippines (Heaney et al. 1989), Madagascar (Ryan et al. n.d.), New Guinea (Flannery and Seri 1975), and probably on any mountain large enough to support a montane fauna. This level is usually known as lower montane forest. I conjecture that the change in structure of the ground surface organic layer at this elevation creates a trophic opportunity for new small-mammal foraging guilds (i.e., under moss) while eliminating the substrate for lowland foraging modes. A thousand meters higher, the moss layer becomes extremely thick on trunks and branches, creating yet another, more arboreal, foraging substrate.

The lowest home ranges of *T. montana* saddle the edge of the mist zone on Mount Kinabalu at Poring. On our high study plot three species of montane squirrels also appeared which did not occur at 800 m on our low plot: *Dremomys everetti, Callosciurus baluensis,* and *Sundasciurus jentinki.* A fourth, *Exilisciurus whiteheadi,* just descended to the top of the lower plot (one sighting). At least two squirrel species from the low plot had dropped out at the level of *T. montana* (*Callosciurus prevosti,* perhaps *C. notatus,* and *Exilisciurus exilis*). Elevational changes in habitat thus affected some squirrels at the same height that they affected treeshrews, but whereas the entire treeshrew fauna turned over near this level, with four lowland species being replaced by two montane ones, at least one squirrel, *Ratufa affinis,* ranged high into the cloud forest.

TREESHREWS IN LOGGED FOREST AT DANUM VALLEY

A major objective of research at Danum Valley has been to study the consequences of selective logging in the surrounding concessions, with the forest of the conservation area serving as the baseline control (Marsh

TABLE 4.4. A, Comparison of captures on unlogged study plot and selectively logged forest at Danum Valley. Total numbers of individuals and young ( ) treeshrews captured. Logged forest trapline in May 1991 had 31 traps; in July, 27 traps. The traplines in the two forests had different configurations and trap numbers, but each had the same basic length (500 m). One or two *T. longipes* in the unlogged plot were trap-shy, so the known population was underrepresented by captures. B, Results of trapping in logged forest and tree plantations by Stuebing and Gasis (1989); each tree plantation had two traplines, and captures from both are added.

Month/Area	T. minor	T. gracilis	T. longipes	T. tana	Total
		A. Present Study			
May					
Unlogged	2 (0)	5 (3)	3 (1)	5 (0)	15
Logged	0	2 (1)	6 (3)	4 (0)	12
September					
Unlogged	1 (0)	1 (0)	2 (1)	6 (1)	10
Logged	0	3 (1)	3 (1)	4 (1)	10
		B. Stuebing and Gasis 1989			
Logged	0	2	5	4	
Eucalyptus	2	0	3	6	
Gmelina	1	0	6	0	
Albizia	2	0	19	3	
Albizia/cacao	0	0	0	0	

and Greer 1992). To get a brief view of treeshrew populations in logged forest, I set a trapline in a "coupe" that had been selectively logged from primary forest in 1987. The trapline on this logged plot was only about 500 m from the nearest unlogged forest, but it was isolated from it by the Segama River and the road (treeshrews are loath to cross wide roads and are rarely seen on them) and extensive logged forest on three sides. Coincidentally, the same transect was used for studies of dung beetles (Holloway, Kirk-Spriggs, and Chey Vun Khen 1992) and fruiting phenology and avian frugivores (Zakariah Hussein pers. com.). The logged forest transect was trapped twice, immediately before the May and September monthly trappings of the main study area.

The tiny trapping sample (table 4.4) shows no evident difference between the logged or unlogged plots for the three terrestrial species, in either relative numbers of adults of each species or rough densities. These species on both plots also appeared to be breeding in the same months. However, the arboreal *T. minor* was not trapped in the logged plot, nor did I see it in several walks around the plot during which I saw the other three species. Because lesser treeshrews are the most readily spotted of all species, this indicates that the population of *T. minor* was much reduced in the logged forest at this time. I also did not see lesser treeshrews during two walks in another logged forest by the Bol River near Danum Valley, although I did see this species in a secondary roadside thicket. I have no data to evaluate the effect of logging on pentail treeshrews, but I saw one in dense roadside shrubs, so they are not restricted to mature forests.

DISCUSSION OF LOGGED FOREST

Stuebing and Gasis (1989) compared small mammal populations in logged forest and in monospecific tree plantations of *Eucalyptus deglupta, Gmelina arborea, Albizia falcataria,* and *A. falcataria* underplanted with cacao (*Theobroma cacao*) within a single large plantation development project in Sabah called Sabah Softwoods. Their trapping methods (bait, trap interval, length of lines) were virtually the same as mine, and their results for treeshrews (table 4.4) can be added to mine to give better insight into interspecific differences in treeshrew habitat requirements. The forest of the plantations had been logged seven years previously, and the tree plantations and their undergrowth were of that age, except that the cacao plantation had been kept clean of undergrowth. The undergrowth vegetation was "luxuriant" in the *Eucalyptus* and *Albizia* plots and sparse in the *Gmelina* stand, due to differences in the canopy density, which was highest in *Gmelina* and lowest in *Albizia,* and perhaps also to the alleopathy of *Gmelina*. Several points emerge from these data.

First, their captures in logged forest were similar in species composition to those at Danum Valley, with *T. minor* absent and the same ratios of the other species. This strongly reinforces the evidence that lesser treeshrews are eliminated or greatly reduced by logging practices that leave other species apparently unaffected. In contrast, *T. minor* is present in all tree plantations except cacao. Second, *Tupaia gracilis,* the rarest species, is absent from all tree plantations. Third, *T. tana* is absent from the two plantations stated to have the lowest undergrowth densities.

Fourth, *T. longipes* is absent only from the one plantation with no undergrowth and survives in one habitat where all other species appear to be extinct. Moreover, *T. longipes* has an apparently threefold population increase in the habitat where undergrowth is most dense.

The only arboreal *Tupaia*, *T. minor*, is also the only species adversely affected by selective logging. It seems likely that this is a direct result of the destruction of the forest canopy: Johns (1992) found (at Ulu Segama, near Danum) that when about 7 percent of the trees were removed commercially, up to 73 percent of the trees in a plot were damaged. After logging almost no closed canopy remains. That lesser treeshrews were present (but rare) in the monospecific tree plantations (Stuebing and Gasis 1989) reinforces this view, because these plantations have a uniform, closed canopy but are otherwise totally different floristically from the native forest.

Parallel to the finding that terrestrial treeshrews had similar populations in logged and unlogged forests, a comparison of invertebrate litter fauna in the same two forests at Danum Valley showed only a slight decrease in arthropods and annelid worms between the two habitats (Burghouts et al. 1992).

The causes of differences in populations in different forest types among the three terrestrial *Tupaia* are unclear. Hypotheses to explain these differences rely on ecological data presented in the next few chapters, so I defer speculation to chapter 12.

APPENDIX: METHODS

Throughout the study I recorded every treeshrew that I saw and its height in the forest. Initially I had to discover which species were present in each study area and learn to recognize them in the field. Some species were common or conspicuous and easy to spot, while others were rare or difficult to identify with certainty without a good view with binoculars. I was expecting to find an additional species, *Tupaia dorsalis*, at Danum Valley, because it had been reported to have been seen on the study area (F. Lambert pers. com.; Danum logbooks). During the first month of my study at Danum, I twice thought I caught a brief, uncertain glimpse of one. However, not only did I never capture one during thirteen months of trapping and never identify one among the approximately five hundred positive field identifications of species; but I also never thought I saw one at the end of the study, when I was most adept at field recognition of treeshrews. I therefore conclude that my early sightings of "*T. dorsalis*" were most likely of juvenile *T. tana*, which were in great numbers at that time, and I look with skepticism on field identifications of this species at Danum (e.g., Anon. 1993) until one is vouchered with a specimen.

Observational Biases. Most of the canopy is effectively screened from view at night by reflections from lower layers of leaves, and most sightings of pentails were of radio-tagged animals that were in the lower layers. High animals are therefore likely to be underrepresented in the data. Likewise, it is also impossible to see by day beyond the screening subcanopy into the canopy above (see chap. 2), or into dense liana tangles, although there is always much greater visibility in daytime than at night.

CHAPTER 5
Diet and Foraging Behavior

Its diet is the most fundamental ecological attribute of an animal. Virtually every feature of a species' biology, from morphology to life history, is linked to what it eats and where, when and how nutrition is acquired. The diet of treeshrews was from earliest reports correctly known to consist of fruit and insects (Cantor 1846). However, this knowledge alone does not distinguish treeshrews from other rainforest mammals, almost all of which also eat fruit and insects (Emmons 1995), nor does it define the ecology of individual species. Below I describe the nature of fruit- and insect-eating by treeshrews, along with the behavioral characteristics of foraging for each food type. I discuss how the diet is related to morphology and review the possible competitive interactions that might occur for different food types both among treeshrew species and among treeshrews and other vertebrates in their community. With this information we can start to approach the question, What is a treeshrew?

FRUGIVORY

CHARACTERISTICS OF FRUIT SPECIES EATEN

As soon as we began to follow treeshrews by radio-tracking at Poring, we discovered that they often spent much of their day at fruit trees (Emmons 1991). I was able to watch lesser treeshrews feeding on fruit, sometimes for hours, but for other species fruit tree use was inferred largely

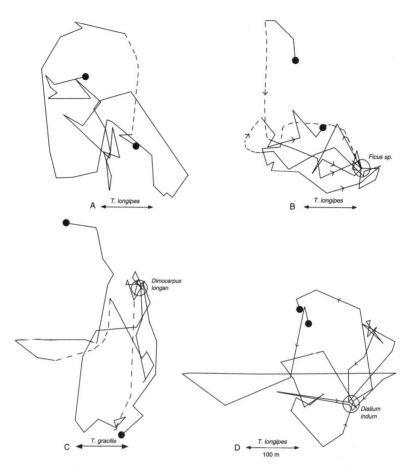

Fig. 5.1. Influence of fruit trees on treeshrew movements. A and B, movements of *T. longipes* female F86 on sequential days. A, without a known fruit tree (26 Mar. 1991, 44 locations, 2,137 m traveled); and B, feeding at a small fig (*Ficus*) tree (circle), to which she returns five times (27 Mar., 48 locations, 2,189 m traveled). C, movements of *T. gracilis* female F67 when feeding on fruit of a mata kuching (*Dimocarpus longan*), which she visits twice (26 Sept. 1990, 54 locations, 2,240 m traveled). D, movements of *T. longipes* male M64 when foraging at a tree of *Dialium indum*, which he visits four times (1 Dec. 1990, 41 locations, 2,512 m traveled). Dashed lines represent approximate routes where radio-tracking data were imprecise.

from movements and rare, brief glimpses. To feed on fruit, a treeshrew might either return to a fruit tree at short intervals throughout the day, again and again heading for it in a beeline, or spend a long time near it once or more times (fig. 5.1). When a treeshrew focused activity in a small area to which it returned repeatedly, I tried to find the source of

TABLE 5.1. List of fruit species recorded as eaten by treeshrews. Site: D = Danum Valley; P = Poring.

Fruit Species	Site	Treeshrew	Size (cm)	Color
Alangiaceae				
Alangium ebenaceum	D	Tg, Tl, Tt	1.5 × 2.0	green
Annonaceae				
Polyalthia sumatrana	D	Tl, Tt	2.8 × 1.2	purple
Grossulariaceae				
Polyosoma integrifolia?	D	Tl, Tt	2.5 × 1.0	green
Leguminosae				
Dialium indum	D	Tl, Tt	2.7 × 2.0	brown
Meliaceae				
Lansium domesticum	D	Tl	6.5	yellow
Moraceae				
Ficus cf. *subulata* iv-2	P	Tl, Tg	2.0	orange/yellow
Ficus cf. *sundaica* ii-7	P	Tmi	1.5	red/orange
Ficus cf. *sinuata*				
or *parietalis* A660	P	Tl	0.8	
Ficus cf. *sumatrana* A500	P	Tg	0.6	
Ficus cf. *benjamina* C-60	P	Tl, Tg, Tm	0.8	red/orange
Ficus sp. C-40	P	Tl	0.5 × 0.7	
Ficus sp. iii-150	P	Tl	1.5	yellow
Ficus sp. A N1W250	D	Tl	2.5 × 1.8	
Ficus sp. Tiny vine				
N0W025	D	Tl, Tt, Tm	1.0	red
Ficus sp. B N2W225	D	Tl, Tg	1.4 × 1.2	
Myrsinaceae				
Sp.	P	Tmi	<0.5	red/purple
Rafflesiaceae				
Rafflesia keithii	P	Tt	14	brown
Sapindaceae				
Dimocarpus longan	D	Tt, Tl, Tg	1.5	brown
Sp.	P	Tmo	0.6	orange aril
Sapotaceae				
Payena acuminata	D	Tt	3.4 × 1.8	green
Cf. *Madhuca* sp.	P	Tt, Tl	3.4 × 1.4	green
Vitaceae				
Parthenocissus sp.	D	Tmi	1.8 × 2.0	yellow

interest by searching the site after the animal had left. This procedure was only partially successful, and the total list of fruit species identified in treeshrew diets was small and probably biased toward the larger, more prolific, and more easily detected sources (table 5.1). Many other fruit species are certainly included in treeshrew diets, but nonetheless the species on the list share attributes that illustrate the nature of treeshrew frugivory.

At Poring figs (*Ficus* spp.) so completely dominated the list of fruits eaten by treeshrews (64%) that tupaiids might have been thought to be fig specialists. However, the next year at Danum figs were only 27 percent of species known to be eaten, and it was clear that fruiting fig trees were simply much more numerous on the study area at Poring than at Danum Valley, where other fruits were eaten more often. In both areas suitable fig trees commanded intense interest from all species of treeshrews, in terms of both hours spent at them and the number of daily visits, so that figs can be said to be among the most preferred fruit (see chap. 8 for data on time spent at fruit trees). Some other fruits elicited only brief visits, perhaps because fruit fell sporadically and little was available at one time or because the animals did not much like them.

Typical fruits eaten by treeshrews were small, 0.5 to 2.0 cm in smallest dimension (table 5.1, fig. 5.2), with soft edible parts (mesocarp, aril, or syconium), and ranged from juicy to dry and to astringent, sweet (most), or quite tasteless (to me). About half were red, yellow, or orange, colors favored by monkeys and birds; two were purple, as favored by birds; and the rest were green or brown, typical of nonprimate mammal dispersed species (Gautier-Hion et al. 1985). Treeshrew fruits thus varied in probable disperser targets.

The sizes of fruit species eaten by treeshrews mirror the most common size classes in the environment (fig. 5.2) but exclude the larger sizes. The single very large fruit identified as eaten was that of the parasitic plant *Rafflesia keithii*. This large, sessile, indehiscent, terrestrial fruit grows from an underground stem and has a soft, oily, strong-smelling, custardlike pulp filled with thousands of tiny seeds. We observed *T. tana* and a squirrel (*Callosciurus notatus*) bite off the pulp of an opened fruit over the course of several hours (Emmons, Nais, and Biun 1991). This strange fruit, with its odd dispersal mechanism, was the only fruit species identified for which treeshrews (with squirrels) are likely to be principal dispersers.

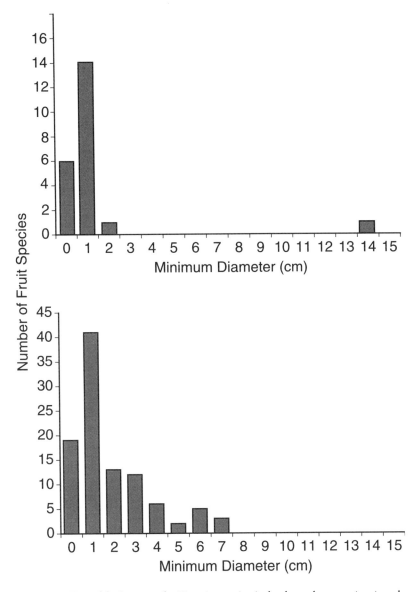

Fig. 5.2. Size of fruits eaten by *Tupaia* species in both study areas (top) and size of all fruits collected in transects at Danum Valley (bottom).

GENERIC DIFFERENCES IN FRUGIVORY

All *Tupaia* species ate fruit and were readily attracted to ripe bananas, on which they gorged themselves when they could, but the other two genera of Bornean treeshrews appeared to be less frugivorous. The smooth-tailed treeshrew, *Dendrogale melanura,* a montane species, cannot be captured in traps baited with fruit (Alim Biun pers. com.). Pentail treeshrews likewise showed scarcely any interest in banana bait, but they did appear to be slightly frugivorous. The sparse evidence for pentail frugivory is summarized below.

Pentails were common on the Danum Valley study area (see chap. 8), but they comprised only 5 of 420 treeshrew captures during the project. Once caught, they ate the banana in the trap, and two of only three scats of this species that were collected from traps for dietary analysis included the remains of fruits other than bait. Captives are said to readily eat fruits (Gould 1978; Lim 1967). During 199.95 h of active foraging by all pentails followed by radio-tracking, *P. lowii* F163 spent 40 min in a tree that supported a fruiting liana of an Annonaceae (cf. *Artobotrys* sp.), from which fruit fragments were falling; but this tree and liana were also popular with palm civets (*Paradoxurus hermaphroditus*), and I could not see into the fruiting canopy to determine what was eating the fruit. During a torrential rain, F163 spent 2.6 h in a giant fruiting fig next to her nest tree, but again her activities there could not be seen. On another night pentail F96 visited a fruiting *Dialium indum* and also a *Polyalthia sumatrana,* both highly favored fruits of *Tupaia,* but she did not linger in either. The degree of frugivory in wild pentails is thus unclear, but they seemed to eat fruit at least sporadically, perhaps often, although without sufficient attraction to be lured by it into traps or to reveal distinctive fruit-feeding behavior during radiotelemetry.

The three terrestrial lowland *Tupaia* species overlapped broadly in the fruit species they were known to eat (see table 5.1). When *Tupaia* of different species were simultaneously radio-collared on the same terrain, they usually visited the same fruit trees. *T. minor* alone was recorded at certain fruits, but conditions of observation usually did not permit viewing the ground under these, and my presence watching the tree would likely have inhibited visits by the sharp-eyed ground-foraging species. Further study is needed to establish whether differences in fruit preference exist among *Tupaia,* but my observations thus far indicate that such differences are not likely.

TABLE 5.2. Fruit transit time in minutes for two species of *Tupaia* in captivity.

Treeshrew (N)	First Marker		Last Marker	
	Mean	Range	Mean	Range
T. minor (7, 4)	20 ± 5	13–29	45 ± 7	38–54
T. tana (5)	57 ± 13	38–73	163 ± 52	118–248

NOTE: At the start of the day's active period, treeshrews were fed one bite of banana containing marker dye, followed by their normal diet *ad lib*. Times are to first and last appearance of marker dye in the feces. Last appearance includes only strongly marked material; faint traces of dye persist beyond that time. N = number of independent trials. (Emmons 1991; experimental details given therein.)

FRUIT FORAGING STRATEGY

FRUIT PROCESSING: TRANSIT TIME When I first captured *Tupaia* species, I noticed that individuals who had been only a short time in a trap were already defecating the banana bait. This led me to carry out some experiments between the two field seasons, at the National Zoological Park in Washington, D.C., to measure the passage times of fruit eaten by *T. minor* and *T. tana* (Emmons 1991). These were followed by more detailed experiments of both passage time and assimilation efficiency of insects (Roberts et al. pers. com.). These tests showed that, for fruit, *Tupaia* have among the most rapid transit times known in mammals, equivalent to those of fruit bats (table 5.2; Emmons 1991). The shortest passage time in seven trials on *T. minor* was only 13 min between eating dye-marked fruit and first defecation of a marker, and that of *T. tana* was 38 min.

Transit time within a taxon of mammals increases with body size, because it is evidently a function of the absolute, rather than relative, length of intestine that the food must traverse (Chivers and Hladik 1980). Predictably, then, because it is a larger animal, *T. tana* has a longer transit time than *T. minor* (table 5.2). The simple intestinal gross morphology of treeshrews is very similar to that of fruit bats (chap. 2) and is doubtless causally related to the short fruit passage times of both groups.

FRUIT PROCESSING: FEEDING BEHAVIOR Shortly before I left Poring I experienced one of the most exciting moments of discovery that year, for it afforded a completely new insight into treeshrew ecology.

20/8/89; 0928 (seen from a blind): A *T. tana* sits on a fallen branch on the ground, eating a green fruit [Sapotaceae: *Payena* or *Madhuca* sp.], holding it

Fig. 5.3. Fruit of Sapotaceae (cf. *Payena*) and wads of fiber spit out by *T. longipes* on 20 August 1989.

in the forepaws like a squirrel. 1045 h: A *T. longipes* picks up one of the same fruits from the ground, climbs up on a fallen log and eats, sitting. It chews and spits out wads of sucked fiber like a bat [see fig. 5.3].

This fiber-spitting behavior is well known for both New World (Microchiroptera; Phyllostomidae) and Old World (Megachiroptera) frugivorous bats. I later saw this behavior of ejecting indigestible seeds and parts of fruits in the captive colony of *T. tana* at the National Zoo (Emmons 1991), but from watching these animals it was unclear to what extent large treeshrews masticate and suck fruit tissue before spitting out unwanted pieces. However, there is independent evidence from a study of masticatory movements in *T. glis* (Fish and Mendel 1982) that when eating fruit this species presses the food against the palate, which is concave and deeply ridged (Martin 1968), with the same tongue movement pattern used by flying foxes, which also have deeply ridged, concave palates. Also like fruit-eating bats, treeshrews salivate abundantly. Copious saliva may aid in the rapid extraction of soluble nutrients from masticated fruit. This whole set of behaviors is thus just like those found

Diet and Foraging Behavior

in specialized fruit bats; but in bats the fruit processing has a more complex associated morphology (specialized teeth, cheek pouches to hold fruit, etc.) and is probably more efficient at extracting nutrients. To date, only bats and treeshrews are known to share this method of eating fruits.

Treeshrews of the genus *Tupaia* have a suite of morphological and behavioral traits associated with fruit processing, which together result in ingestion of only the most quickly digestible fractions of fruits, which pass rapidly through the gut. It is unlikely that any digestion or fermentation of structural plant carbohydrates can occur, as there is neither enough time during passage nor any digestive tract site for it. This implies that fruit contributes only its most readily assimilated nutrients to treeshrew diets. It is not known whether *Ptilocercus* or other treeshrew genera share these traits.

TEMPORAL PATTERN OF FRUIT-EATING

The physiological behavior outlined above relates to the temporal patterns of fruit-eating observed in the field.

> 0745: *Tupaia minor* F112 is foraging in a fruit tree [Myrsinaceae]. She eats a ripe, red fruit, holding it in her paws, then leaves tree at 0746. . . . 0816: She goes rapidly straight back to the fruit tree, arriving at 0823. 0827–0847: She rests, crouched on a branch. 0847–0858: She forages, eating fruit. 0900: She goes back and rests on the same spot. 0910: She grooms several parts of body and tail, then curls up and rests again. 0913: She gets up and forages, then leaves the tree at 0926.

The alternation of short feeding bouts with short resting bouts was characteristic of *T. minor* whenever I watched them feeding on any species of fruit. The interval between feeding bouts approximates the transit time of fruit through the gut of lesser treeshrews (see table 5.2), so that it seems that the animals are filling up with fruit repeatedly, as often and quickly as it can be transited through the digestive system.

At Danum Valley a huge *mata kuching* tree (*Dimocarpus longan*, a delicious, juicy fruit much exploited by people as well as by most wild frugivores) massively fruited within the ranges of radio-collared treeshrews of three species. To quantify visits to a major fruit source, on 28 and 29 September 1990, I monitored this tree continuously from afar, with a radio receiver, from before dawn until heavy rains stopped all activity in late afternoon. I also followed one of the treeshrews (*T. gracilis* F67) continuously from dawn to dusk during four other days of the same week. The results show that like lesser treeshrews these three other

TABLE 5.3. Time spent per day by females of *T. tana* (F65), *T. longipes* (F56), and *T. gracilis* (F67) at a fruiting mata kuching (*Dimocarpus longan*).

Treeshrew	Date	No. of Visits	Hours at Tree
T. tana			
F65	28 September	3	1.60
	29 September	0	0
T. longipes			
F56	28 September	2	0.67
	29 September	6	1.15
T. gracilis			
F67	26 September	2	2.60
	28 September	3	1.03
	29 September	5	0.72
	1 October	4	0.48
	2 October	3	4.84
	3 October	2	0.93
Total		30	

NOTE: The *T. tana* and *T. longipes* were monitored only on 28–29 September 1990. The *T. gracilis* was followed continuously on the four other days.

Tupaia species often spent long periods next to a fruit tree, during which they probably fed more than once, or made multiple visits to it each day (table 5.3). These observations were at the peak of the fruit masting episode, when many other trees were fruiting and the treeshrews were all likely to be eating other fruit species as well.

Although I was unable to see them eating, continuous radio-tracking showed that the pattern in which the three terrestrial species visited fruit trees was similar to that of lesser treeshrews I could watch, and I surmise that they feed similarly in bouts, perhaps with longer intervals between meals, commensurate with their longer passage times. Many other radio-tracking episodes, and feeding behavior in captivity, support this view.

SEASONAL FRUIT ABUNDANCE

The amount of fruitfall registered in our monthly transects on the study plot at Danum (fig. 5.4; see also chap. 3) showed large seasonal variation. The number of fruit species recorded per month varied by a factor of nearly eight; however, the amplitude during that year was dominated

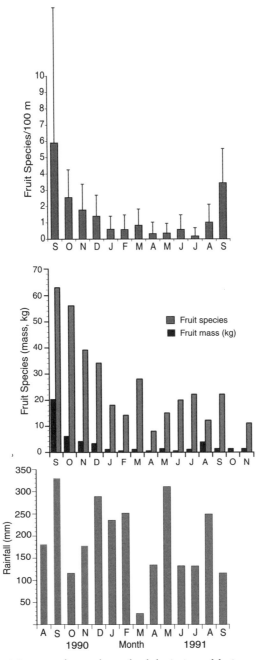

Fig. 5.4. Mean number and standard deviation of fruit species collected monthly per 100 m of trail on the Danum Valley study area (top); fruit species (light shading) and total weight of fruit (dark shading) collected monthly on a transect of 4.5 km of trails (middle); and rainfall at the research center during those months (bottom). There was no fruit collection in October 1991.

TABLE 5.4. Months in 1990 and 1991 during which fruit species were eaten by treeshrews at Danum Valley.

Species	A	S	O	N	D	J	F	M	A	M	J	J	A
Payena acuminata	X												
Dimocarpus longan		X	X										
Parthenocissus sp.			X										
Lansium domesticum			X										
Polyosoma integrifolia?			X										
Polyalthia sumatrana			X	X									
Dialium indum					X								
Ficus sp. tiny vine								X					
Ficus sp.										X			
Alangium ebenaceum												X	X
Ficus sp. N250W2												X	
Ficus sp. N1W250													X

NOTE: No observations were made in January and February, and full-time radio-tracking, when best observations were made, began 17 September 1990.

by the unusual masting phenomenon, and it would be much less in a nonmasting year. We did not measure the actual density of any individual fruit species, but we recorded the incidence of each species in each of the forty-five 100 m segments of the transect. This should be directly related to the number of fruiting points. These data (fig. 5.4 top) show a curious departure from the overall fruiting pattern (fig. 5.4 middle) in that the May to July surge in fruit species number was not mirrored by an increase in the incidence of fruit species per 100 m, which was continuously low from January to July, with a minor increase only in March. For almost all months, the standard deviation of the number of fruit species per 100 m exceeded the mean, demonstrating that fruit was highly unevenly distributed in space.

A summary of months in which fruit species were recorded as eaten by treeshrews at Danum (table 5.4) shows that only in October of the masting peak was more than one or two species registered; I saw no treeshrews eating fruits in May or June. During particular months at Danum Valley, certain fruit species seemed to dominate the interest of treeshrews, to the point of being apparent "keystone" resources. These dominant fruit species included *Payena acuminata* (or similar Sapotaceae) in August; *Polyalthia sumatrana* in October and November; *Dialium indum* in December; and *Alangium ebenaceum* in July and August. At Poring species of *Ficus* as a group were the major and almost only dominant fruits in

Diet and Foraging Behavior

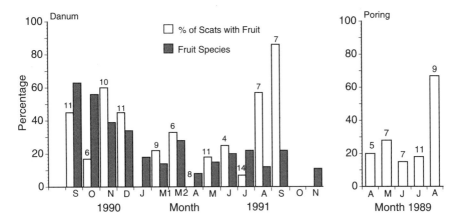

Fig. 5.5. The percentage of treeshrew scats (all species) collected on the Danum study area that contained fruit other than bait, by month (left figure). Total 104 samples, N for each month above the bar, no data for January, 0 percent fruit in April; February set collected in early March as in figure; compared with the total number of fruit species collected each month in the fruit transect. The percentage of scat samples with fruit per month at Poring, total 39 samples, including *T. montana* (right figure).

the diet during the few study months. More than one species of treeshrew generally were recorded feeding on each of these key fruits, but radio-tracking samples, when fruit-eating was usually found, did not always include every treeshrew species during each month, so lack of data does not necessarily mean that a treeshrew species did not eat a particular fruit.

The general annual pattern of fruitfall (fig. 5.4) will only reflect the fruit resource available to treeshrews if the species they eat have a phenology resembling that of the forest as a whole. To evaluate this, the overall presence/absence of fruit (other than bait) in 143 scat samples collected throughout the study can be compared to the general phenology (fig. 5.5). Too few samples are available to analyze the data for single treeshrew species, but since all appeared to eat the same fruits, it seems legitimate to combine them. Considering the small monthly sample sizes, the percentage of treeshrews that had eaten fruit shows a surprisingly exact, almost perfect correspondence with the number of species collected on the transect from September 1990 to June 1991, with only October 1990 departing from the trend. Particularly notable is the complete absence of fruit in scats during the month of minimum fruitfall (April). However, the final two months of the study (August and September) show a sharp deviation from concordance, with fruit eaten often during a period of low fruit abundance. I tentatively interpret this as showing that during these months fruit of interest to treeshrews were proportionally better represented in the fruit species pool than in other months. We will see that this has relevance to reproduction (chap. 9). The 1989 data from treeshrew scats at Poring shows a closely similar, low incidence of frugivory from April to July, followed by a sharp increase in August (fig. 5.5).

INSECTIVORY

FORAGING BEHAVIOR

Each treeshrew species had a distinctive way of searching for invertebrate prey. Olfaction was clearly dominant in many cases of searching and prey capture, but vision was evident in others.

PTILOCERCUS LOWII The most characteristic hunting behavior of pentails was exploring the surfaces of tree trunks, which could be bare but more often had attached lianas. The trunks on which they foraged ranged from slender understory saplings to the largest boles of emergent giants. Pentails also searched up and down branches or liana stems, and I once saw one repeatedly hunting up and down in a hanging mass of thin dead twigs in a dead, leafless vine festoon. One searched the surface of a dead, rotten, standing stump, thoroughly investigating inside an empty woodpecker hole (to which I later climbed to examine). These treeshrews were not often seen in leafy foliage, but when they were, they were in the tops of treelets in the understory or subcanopy. It was not possible to watch pentails for long, or without some disturbance from the light, because they moved rapidly and often avoided the headlight beam (although some individuals became quite habituated). Also, they could not be seen when they were in the canopy. Within these limitations, I would define *P. lowii* as a forager on arboreal woody surfaces—the bark of trunks, stems, and branches—with less attention to the foliage of treelets. A species-typical behavior was to hunt up and down the same tree trunk several times, before scampering quickly on to the next one.

TUPAIA MINOR Lesser treeshrews hunted arthropods by moving steadily through the vegetation, investigating every likely insect hiding place. Favored search sites included hanging masses of dead leaves, rolled dead leaves, dead twigs, holes in dead wood, crevices between stems and trunks, and within root masses of epiphytes such as ferns. These places were examined by poking the nose into them or sniffing along them, evidently hunting out hidden prey by smell. Living foliage was also carefully searched by nosing it, but leaves and other surfaces were often examined by peering intently at them, investigating visually. Dense vine tangles were by far the most popular hunting grounds of lesser treeshrews, and a single, densely vine-enveloped emergent tree could occupy several hours of foraging. Such places doubtless include the highest densities of living and dead foliage, hung-up debris, stems, and crevices that harbor arthropods. As they traveled, lesser treeshrews also searched along the branches and through the foliage of trees. While hunting in living foliage, they progressed on the stems within and generally examined the undersides of the leaves, often craning their necks to study the leaves above. They hunted sheltered from above, inside, or under the leafy surface. They did not linger on bare trunks or stems, or often use the outside surface of the foliage, although I saw one foraging on outer canopy foliage by jumping from branch tip to branch tip. I saw lesser treeshrews feed on large Orthoptera and spiders, but few prey could be identified in the field.

TUPAIA GRACILIS Slender treeshrews also hunted by sniffing and looking. I most often saw them working along at ground level, on the ground surface, on fallen logs, roots, or rotten wood, but they typically searched the living leaves of the undergrowth plants, peering or nosing up to the undersurface of foliage above them, or climbing through the understory shrubs or vines.

> *T. gracilis* F67 seen insect foraging in vines 1–6 m, she disturbed an insect that dropped down and she raced down the vine after it to the ground. A few seconds later I saw her eating a large insect.
> A *T. gracilis* foraging in the lower 2 m of aboveground, dense vines, eats a large, hairy caterpillar, climbing to 1.5 m on a slanting stem to eat. She rubs it once on the branch, holding it in the mouth.

I saw slender treeshrews pick up prey from the ground surface, as well as from shrubs and a rotten stump, and I saw them sniff dead leaves. Most of the few prey items I saw were large (but large items are easier to see than small ones): one, "almost as large as she (F67), a huge caterpillar or katydid." Slender treeshrews are surface gleaners of the ground and understory plants, with rare forays up to the lower canopy.

TUPAIA MONTANA Too short a time was spent in the habitat of this species to characterize well its hunting behavior, but both sightings and radio-tracking showed strongly terrestrial habits, with a predilection for logs and roots and the cavities beneath these.

TUPAIA LONGIPES The predominant foraging pattern of plain treeshrews was to travel quite quickly along, nose to the substrate: "A *T. longipes* forages rapidly along the ground, nibbling at things on the litter surface" (7 November 1990). I was not able to identify any prey item, as these treeshrews generally picked up small items and continued to travel as they chewed. The most common substrates were the ground and rotten logs, but they often climbed around, to the height of about a meter, in dense, bare snarls of liana stems in the dark recesses below blankets of climbers, or walked along the larger horizontal lianas of such places. Rarely, plain treeshrews foraged on the broad sides of large logs or trunks of live trees or higher in vines (see table 4.1). This species was a surface gleaner of its substrate. I did not see it foraging in foliage or digging in the litter.

TUPAIA TANA Unlike the other two lowland terrestrial species, large treeshrews typically foraged for items hidden under the surface litter of the decomposing organic layer.

> *Tupaia tana* F58 came quietly sniffing along a liana on the ground, then nosed along the litter. She dug a hole in the litter, alternately digging and sticking her nose in and sniffing. She dug only into the surface layer of a place with a thick leaf mat [5 November 1990].

Large treeshrews would walk very slowly along, periodically pushing the nose into the litter and digging when they discovered a morsel of interest. Mesic sites, such as streamsides, dark tunnels under dense vinefalls, wet, rotten logs, and cavities beneath logs, upended stumps, or brushpiles were preferred foci of activity. The foraging speed of this species was strikingly slower than that of *T. longipes* or *T. gracilis*, especially after rain (see chap. 8). Unlike the other species, large treeshrews would often spend many hours in a tiny sector of their home range and return to it day after day. When I searched these areas I never found fruit sources, so I assume that the animal prey of this species was concentrated in particular microhabitats. All of these intensively used "foraging" sites (a supposition) of *T. tana* were on the banks of streams or rivers. Large treeshrews also picked up prey from the surface of the substrate and broke into rotten wood (Alim Biun pers. com.). They did not seem to forage on foliage or up in vines or shrubs, although they jumped up onto these to scan the surroundings.

When excavating prey, large treeshrews seemed to dig only shallowly into the top layers of decomposing litter. Digging behavior is strongly programmed into *T. tana*: captive individuals at the National Zoo that are kept in large cages floored with a deep layer of bark chips, though fat and satiated with food, spend much time walking slowly, probing into the substrate with the nose every few steps, exactly as they do in the wild.

GENERAL TYPES OF INVERTEBRATE PREY

The prey eaten by treeshrews was determined by microscopic examination of feces collected during monthly trapping. Warren Steiner, a Museum Specialist in Entomology at the Smithsonian Institution, undertook this tedious but exacting task, and the results below are products of his great patience and skill. Appendix IV is a complete list of identified taxa.

The many intrinsic sources of variability associated with treeshrew fecal samples make it difficult to quantify the contents in percentages of total diet. One difficulty stems from the short passage time: a fecal sample probably contains only a fraction of a day's foraging activity, so its contents may change quickly from hour to hour and depend on the time of day that the animal was trapped. Different food types have different passage times, so they have different likelihoods of being recorded in a given time frame. Some fruits might be eliminated in scats in less than an hour, insects in several hours (M. Roberts pers. com.), whereas abundant tiny items such as lepidopteran scales or worm setae from a single food item may trickle out over longer periods. Much of the fecal material is an unidentifiable matrix, which is unlikely to represent a proportional sample of the identifiable, because food items intrinsically differ in digestibility. Finally, the amount of material derived from wild meals per scat varied greatly, from one or two ant heads (the rest bait) to large quantities of entirely prey items. A treeshrew may have been in a trap from minutes to hours.

For these reasons, I evaluate the representation of kinds of invertebrates in the diets simply from the number of samples in which items occurred (percent occurrence, or frequency). The number of individuals and species of a given prey taxon that could be recognized in each sample was counted, or, for ants and termites, classed in general abundance ratings ($1 = < 10$; $2 = 10–20$; $3 = 20–30$; $4 = >30$). The numbers of individuals of each prey type eaten gives an idea of the relative degree to which an animal concentrates or specializes on particular kinds of items.

The analysis of scats shows prominent divergences between treeshrew

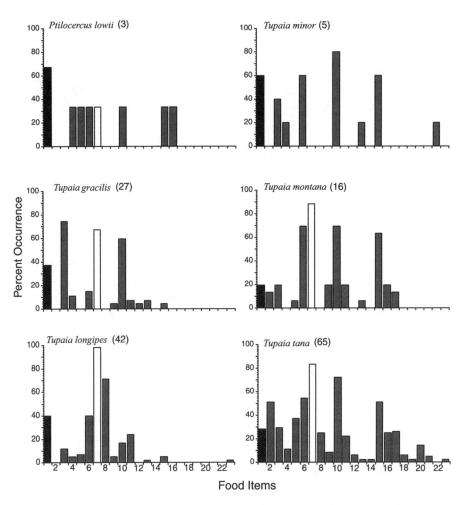

Fig. 5.6. Percent occurrence of food items in treeshrew diets × axis units: 1 = fruit; 2 = earthworms (setae present); 3 = Lepidoptera (larvae); 4 = Lepidoptera (adults); 5 = Coleoptera (larvae); 6 = Coleoptera (adults); 7 = ants; 8 = termites; 9 = wasps; 10 = Orthoptera; 11 = Hemiptera; 12 = Diptera; 13 = Homoptera; 14 = Odonata; 15 = spiders; 16 = millipedes; 17 = centipedes; 18 = uropygids; 19 = phalangids; 20 = scorpions; 21 = crabs/crayfish; 22 = molluscs; 23 = lizards.

species in their choice of invertebrate prey (fig. 5.6; tables 5.5, 5.6). One of the most surprising findings was that, despite what appeared to be generalized or nonspecific searching behavior, most species seemed to specialize to the extent that invertebrates of only three or four orders strongly dominated in the samples (each found in 60–98% of the scats) while all other orders appeared infrequently (5–10%).

TABLE 5.5. Occurence of food items in diet. Number of scats in which item found (N) and percent of total number of scats in which item found.

Food	T. gracilis (N)	%	T. montana (N)	%	T. longipes (N)	%	T. minor (N)	%	T. tana (N)	%	P. louvii (N)	%
Fruit	(10)	37	(3)	19	(17)	40	(3)	60	(18)	28	(2)	67
Earthworm			(2)	13					(33)	51		
Caterpillar	(20)	74	(3)	19	(5)	12	(2)	40	(20)	31		
Lepidoptera adult	(3)	11			(2)	5	(1)	20	(6)	9	(1)	33
Coleoptera larva			(1)	6	(3)	7			(24)	37	(1)	33
Coleoptera adult	(4)	15	(11)	69	(17)	40	(3)	60	(35)	54	(1)	33
Ants	(18)	67	(14)	88	(41)	98			(54)	83	(1)	33
Isoptera					(30)	71			(16)	25		
Hymenoptera	(1)	4	(3)	19	(2)	5			(5)	8		
Orthoptera	(16)	59	(11)	69	(7)	17	(4)	80	(47)	72	(1)	33
Hemiptera	(2)	7	(3)	19	(10)	24			(14)	22		
Diptera	(1)	4							(4)	6		
Homoptera	(2)	7	(1)	6	(1)	2	(1)	20	(1)	2		
Odonata									(1)	2		
Aranae	(1)	4	(10)	63	(2)	5	(3)	60	(33)	51	(1)	33
Diplopoda			(3)	19					(16)	25	(1)	33
Chilopoda			(2)	13					(17)	26		
Uropygida									(4)	6		
Phalangida									(1)	2		
Scorpions									(9)	14		
Decapoda									(3)	5		
Mollusca					(1)	2						
Lizard					(1)	2			(1)	2		

TABLE 5.6. The five major invertebrate taxa eaten by each treeshrew species, in percent ocurrence ( ) in scats.

T. minor	T. gracilis	T. longipes	T. tana	T. montana
Orthoptera (80)	**Caterpillars (74)**	**Ants (98)**	**Ants (83)**	**Ants (88)**
Coleoptera adults (60)	**Ants (67)**	**Termites (71)**	**Orthoptera (72)**	Coleoptera (69)
Aranae (60)	**Orthoptera (59)**	Coleoptera (40)	**Coleoptera adults (54)**	Orthoptera (69)
Caterpillars (40)	Coleoptera (15)	Hemiptera (24)	**Earthworms (51)**	Aranae (63)
Lepidoptera, adults (20)	Lepidoptera, adults (11)	Orthoptera (17)	**Aranae (51)**	Caterpillars (19)
Homoptera (20)				Hemiptera (19)
				Hymenoptera, non-ant (19)
				Millipedes (19)

NOTE: Taxa dominant found in more than 50% of scats, in boldface.

PTILOCERCUS LOWII Only three scats were collected from pentails. These included twelve prey items of seven general types (table 5.5, Appendix IV). Beetles, with four individuals, predominated in this tiny sample, which showed only ingestion of a broad range of prey types consistent with an arboreal trunk-foraging mode. This sparse data can be augmented with records from the literature: Lim (1967: 377) reported four *P. lowii* stomachs to contain "chiefly black ants, along with cockroaches, beetles, earwigs, cicadas, and leaf-insects. One contained remains of a young forest gecko, *Gymnodactylus marmorata*."

TUPAIA MINOR Lesser treeshrews were also poorly sampled, with only five scats (feces fell out of arboreal trap sets), including twenty-eight prey of seven types. This small sample nonetheless showed a strong trend: 50 percent of all prey items were Orthoptera of two kinds, crickets and cockroaches; 21 percent were adult beetles; and 10 percent each were caterpillars and spiders. This array again seems to correspond nicely to the observed foraging behavior within arboreal debris, crevices, foliage, and stems. Ants were absent from my sample for this species, but both Davis (1962) and Lim (1967) found ants in *T. minor* stomachs, as well as some of the same arthropods I report.

TUPAIA GRACILIS Slender treeshrews were remarkable for the low diversity of their eighty-six prey, from twenty-six scats, which was dominated by just three types, caterpillars (40% of individuals; Appendix IV), ants (23%) and Orthoptera (21%) (fig. 5.5). The major prey was caterpillars, and 55 percent of the samples that included these contained two or more. Although ants were the second most common prey, these were eaten in small numbers, with 89 percent of occurrences at abundance 1 and 11 percent at 2. A single orthopteran was found 87 percent of the time, two in the other 13 percent of samples. Completely absent from the samples were earthworms, termites (Isopoda), millipedes (Diplopoda), and centipedes (Chilopoda), and there was a single spider (Arachnida, Aranae). Ten general prey types were identified in the samples. The prey found in *T. gracilis* feces correlates well with its observed foraging behavior of surface gleaning, but the dominance of caterpillars points to underleaf scanning as the major foraging surface and mode, a tendency that was not obvious in the field, although I occasionally saw it. The few Orthoptera identified were mainly crickets and cockroaches, both of which are present at all forest levels, but by day these are generally in hidden refuges or cryptically immobile.

TUPAIA LONGIPES Plain treeshrews ate two types of prey in major amounts and three others to a much lesser extent. Ants were the most frequent prey (98% occurrence), and they were also eaten many at a time, with 80 percent of occurrences at abundances of 2 or more and 22 percent at abundance 4. Twelve percent of ant occurrences included broods (eggs, larvae, or cocoons), showing that nests were rifled. Termites were the other main prey of *T. longipes,* and these were likewise eaten in large numbers (at abundance 2 or more in 50% of scats). Beetles (Coleoptera), bugs (Hemiptera), and Orthoptera were the most frequent, but much more scarce, lesser prey. Eleven of the general prey types were found among 220 prey individuals in forty-two plain treeshrew scats (fig. 5.6), and they had eaten no earthworms, millipedes, or centipedes. A reptile tail tip, probably from a lizard, was in one sample.

The prey of this treeshrew would only partly have been predicted from foraging behavior that I was able to observe. The ground, log, and liana-surface nosing and nibbling of tiny items by *T. longipes* conforms with ant eating, or random gleaning of beetles and the eight minor prey types. In particular, the relative absence from the diet of hiding arthropods such as cockroaches and spiders, or foliage insects such as caterpillars, concurs with an observed lack of both foliage gleaning or crevice searching in this species. However, I did not see anything in the field that I would have specifically interpreted as termite hunting. The termites eaten are pale-bodied species, such as *Nasutitermes,* that are not active on the surface by day. To eat these, the treeshrews would have to open their tunnels or nests (but no brood was found in scats). Perhaps termites are concentrated in the dense treefalls and vine thickets favored by plain treeshrews.

TUPAIA MONTANA Montane treeshrews, like plain treeshrews, ate ants most frequently, and they ate many at once (abundance 2–3 67% of the time), but they ate beetles, spiders, and Orthoptera nearly as often (fig. 5.5) and four additional prey types in almost 20 percent of occurrences each. They ate no termites, but they did eat earthworms, millipedes, and centipedes. In the smaller number of montane treeshrew scats examined (16), I found that they ate twelve general prey types (among 99 individuals), or more than either slender or plain treeshrews. Compared to the latter, montane treeshrews had a more diverse and even diet, less skewed toward particular taxa, although still weighted heavily toward four. This is the only treeshrew for which spiders were a major prey (12% of all items but in 63% of all scats).

The prey array of montane treeshrews shows clearly that they seek

Diet and Foraging Behavior

out hidden ground-level prey—some by digging into the litter layer (worms, centipedes), others in crevices (spiders, beetles) or on the surface (ants)—but do little foliage gleaning (few caterpillars). Our few observations agree with this strongly terrestrial mode but alone could not have predicted the arthropod preferences of this species.

TUPAIA TANA   The prey profile of large treeshrews reflects a distinctive feeding strategy (fig. 5.6). Each scat had both more individual arthropods and more general categories of them than did those of other species. Beetles, ants, spiders, and Orthoptera were the most often eaten categories, in common with one or another of the other treeshrews, but several prey types rarely eaten by other treeshrews were important components of large treeshrew scats, including earthworms (51% of scats), centipedes (26%), millipedes (25%), and beetle larvae (37%, mostly elaterid and scarabid). Among the ninety-one individual Orthoptera identified to family, 59 percent are cockroaches and 22 percent are crickets. This was the only species to have eaten arthropods of several other orders or classes, including nineteen scorpions, four uropygids, one phalangid, two crabs, and a crayfish. This eclectic prey range splendidly mirrors the behavior of *T. tana* as seen in the field. Invertebrates gleaned from the substrate surface are mixed with those from beneath the litter surface (worms) or nocturnal species hidden by day under logs, roots, or leaves (centipedes, termites). The beetle larvae may be broken from rotting wood. Many samples for this species included what seemed to be soil. This may have come from the gut contents of ingested earthworms, or else was swallowed while unearthing prey from the soil. The decapods reflect the foraging along streams that was documented by radio-tracking (see chap. 8).

SPECIFIC TYPES OF INVERTEBRATE PREY

Each treeshrew species ate a unique balance of arthropod prey types, but despite this there was a surprising uniformity among the treeshrews in the more specific taxa of invertebrates eaten (Appendix IV). The most intriguing case was that among thirty-one hemipterans that were identified to family, twenty-five (81%) were Reduviidae (plus two coreids and four cydnids). Other strongly dominant prey taxa were lycosids among spiders, scarabs and weevils among beetles, cockroaches and crickets among Orthoptera, and scolopendrids among centipedes. Considering the vast diversity of arthropods in the tropics, relatively few families were identified in the scats. It may be that these are simply the taxa that remain most identifiable following passage through a treeshrew; but it could also be that these are the most numerous taxa where treeshrews forage, and/or those that their foraging modes and sensory skills can most easily detect and capture. Small lizards were the only vertebrates eaten (and possibly a snake tail tip), and it is unlikely that vertebrates are a significant part of the diet.

ABUNDANCE AND SEASONALITY OF INVERTEBRATES

I did not measure the invertebrate prey base during this study, but data on the abundance and phenology of litter invertebrates at Danum Valley were collected in the year before my project during a study of litter decomposition (Burghouts et al. 1992), and some data on numbers of canopy invertebrates are available for another part of Borneo (Stork 1991). These can be used as a rough gauge of whether treeshrews forage randomly or selectively.

A comparison of what treeshrews eat (table 5.5, fig. 5.6) with what is presumably available (fig. 5.7) shows that treeshrews generally eat the most common, nonflying or slow-to-fly large arthropods. Both fast-flying (Diptera, Hymenoptera) and tiny (Collembola, Psocoptera, Thysanoptera, Isopoda, Acarina, etc.) forms are apparently eaten in far fewer than their relative numbers in the environment. Ants are by far the most numerous litter-fauna arthropods (fig. 5.7A), and they are also the most frequent prey in the diets of the three most terrestrial species. Beetles and spiders are likewise dominant both in the habitat and in scat samples. Unlike these, several prey items appear to be taken by particular species of treeshrews in proportions greater than their occurrence. Plain treeshrews seek out termites, and large treeshrews discover cockroaches and crickets with unusual frequency. Unfortunately, studies of the arthropod faunas of understory vegetation, the layer used by *T. gracilis*, seem to have been done only in the New World. One such study in Panama (Greenberg and Gradwohl 1980) found understory leaf arthropods in the following rank orders and numbers: 1, spiders (350); 2, beetles (304); 3, Homoptera (159); 4, Diptera (86); 5, Orthoptera (85); 6, Hemiptera (61); 7, lepidopteran adults (56); 8, lepidopteran larvae (31). Because caterpillars have not been shown to dominate the arthropod faunas of any tropical rainforest level, it is likely that the high frequency of caterpillars in *T. gracilis* scat samples is a result of specialized hunting behavior. The predilection of several treeshrews for reduviid bugs also seems to be the product of more

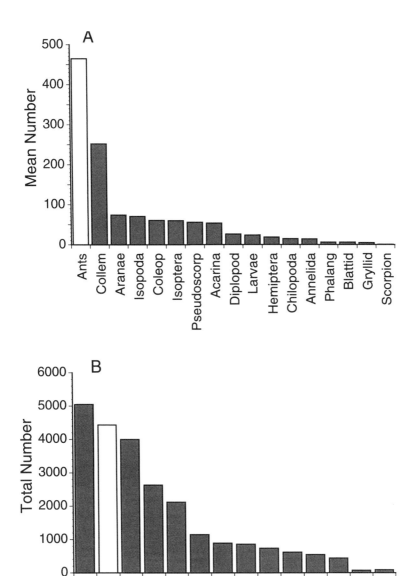

Fig. 5.7. A, average numbers of arthropods in standard samples (25 cm dia x 20 cm deep) of litter collected at Danum Valley (data from Burghouts et al. 1992). B, total numbers of arthropods collected from ten Bornean trees by canopy fogging (data from Stork 1991); most Lepidoptera are larvae (558), and the majority of Orthoptera other than Blattoidea (listed separately) are crickets (Gryllids, 336).

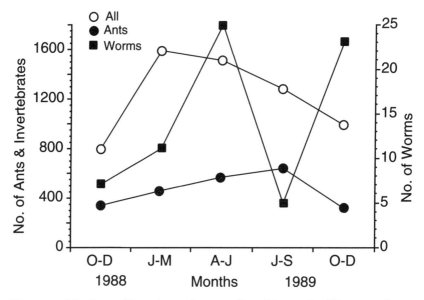

Fig. 5.8. Numbers of litter invertebrates collected in standard litter samples during three-month periods in 1988–89 (Burghouts et al. 1992; T. Burghouts pers. com. of data).

than random gleaning. Stork (1991) found only 27 reduviids among 23,874 arthropods in his canopy sample; and Burghouts et al. (1992) counted only 19 hemipterans in their total of 1,231 litter fauna arthropods at Danum. Reduviids are often associated with termite nests (W. Steiner pers. com.), and possibly *T. longipes* encountered them in especially high numbers (15 identified) while hunting termites.

The pattern of litter arthropod abundance shows about a two- to fivefold seasonal variation in invertebrate numbers (fig. 5.8). The pattern of earthworm numbers generally follows the rainfall pattern for that year, indicating that earthworms are most numerous in the surface litter during wet periods. During periods of drought, they may burrow down out of reach. Arthropods in general had highest numbers from January to March and gradually declined to minimum numbers during October and November, but ants exhibit relatively little seasonal variation (Burghouts et al. 1992). This should make ants a more reliable food source.

DISCUSSION OF FOOD HABITS

Some information on the food habits of treeshrews is reported in the literature (Davis 1962; D'Souza 1972; Harrison 1954; Langham 1982; Lim

1965, 1967; Lyon 1913; Thomas and Wroughton 1909), and only for *Tupaia gracilis* are the first data on the diet published here. The outstanding previous records were those of Davis (1962), who analyzed stomach contents of *T. tana, T. longipes,* and *T. minor,* with results that agree with those we obtained from scat samples. Although frugivory had been noted in treeshrews, tupaiids had generally been supposed to be chiefly insectivorous (Butler 1980).

In frugivory I saw no evident ecological separation between the species, other than the spatial distinction imposed by the arboreality of the two smallest treeshrews. Two species (*P. lowii* and *T. tana*) seemed to be less drawn to fruiting trees than the others (see also chap. 8), but as actual amounts eaten were not measured, the evidence is circumstantial. The seasonal pattern of fruit eating seemed to follow the habitat-wide overall fruiting phenology, with the exception that frugivory seemed to increase in August and September 1991, when overall fruit species availability was low. Fruit eating was facultative, in that during some months treeshrews were not detected eating any. Nonetheless, I will suggest below (chap. 10) that fruit is a most critical portion of the diet.

In contrast to their frugivory, in insectivory the six species showed strong differences in prey composition. These differences arise from distinctly divergent foraging patterns and sites (table 5.7). An unexpected discovery was that *T. montana,* which closely resembles *T. longipes* in size and superficial appearance, does not resemble it in foraging mode and diet but instead is much like *T. tana,* which has morphological adaptations for digging (long claws) and probing under the litter (long snout) that are lacking in the montane species. Perhaps the deep moss substrate of montane forest requires less specialized digging apparatus than the soil of the lowlands. Phylogenetically, mountain treeshrews and large treeshrews are sister taxa well removed from the lineage of plain treeshrews (Han, Sheldon, and Stuebing n.d.), so their dietary similarities reflect a shared evolutionary relationship.

The structuring of feeding roles in treeshrews closely parallels the way that birds segregate into highly specific insect-feeding guilds (e.g., Robinson, Terborgh, and Munn 1990). The similarity in insects eaten (ants, reduviids, lycosids, cockroaches) shows that the basic sensory and prey-catching modalities have remained alike in different treeshrew species, while the foraging sites have diverged, to create a community of morphologically quite similar but ecologically differentiated species. In the following section and chapters we shall see that the foraging mode and

TABLE 5.7. Summary of invertebrate foraging characteristics of the six treeshrew species.

Species	Chief Foraging Site	Dominant or Special Prey
P. lowii	Arboreal: tree trunks and stems, foliage of treelets	Wide variety
T. minor	Arboreal: lianas and living foliage, dead leaves	Orthoptera, Aranae, beetles, caterpillars
T. gracilis	Terrestrial and understory: undersurface of shrub foliage, ground surface, low plants	Caterpillars, Orthoptera, ants
T. montana	Terrestrial: under logs, digs in litter	Ants, Coleoptera, Orthoptera, Aranae, Myriapoda
T. longipes	Terrestrial and surface of logs: ground surface	Ants, termites
T. tana	Terrestrial and logs: digs under litter, logs, and roots	Ants, cockroaches, earthworms, Aranae, Myriapoda, Crustacea

substrate differences among treeshrews carry consequences throughout many other aspects of their lives.

ECOLOGICAL CONSEQUENCES OF THE
DIETARY CONSTRAINTS OF TREESHREWS

The rapid passage time of fruit through the gut of *Tupaia* species (20–60 min depending on animal size) and batlike fruit-processing behaviors, both associated with the simple gross morphology of their digestive tracts, make it likely that only the most readily available (soluble) nutrients are extracted from fruit. The dietary role of fruit may largely be to provide quick and economically acquired calories and nutrients that are scarce in arthropods, particularly calcium. *Tupaia* feed on fruit in short, repeated bouts. Wild treeshrews may fill up again on fruit as soon as the previous batch has passed through the digestive tract. By resting near a tree and feeding at intervals, a treeshrew may achieve maximum intake for minimum effort. An outcome of this feeding mode is that the treeshrews seemingly must eat frequently, with repercussions on the general daily activity patterns (see chap. 7).

The use of fruit by treeshrews can without quibble be called primitive, as it is about the most rudimentary possible, the result of posses-

Diet and Foraging Behavior

sion of a simple digestive morphology. Treeshrews are interlopers on a system of edible fruits and disperser frugivores to which they probably contribute little effective seed dispersal compared to that of the birds, bats, primates, and civets, which feed on the same fruits.

INTERSPECIFIC INTERACTIONS AND FORAGING

COMPETITION FOR FRUIT AMONG TREESHREWS

The treeshrew species seemed to overlap broadly, perhaps completely, in the species of fruit that they ate. Theoretically, they thus compete directly for this resource, although the arboreal and terrestrial species exploit spatially and temporally separated portions of the same fruit supply, of which the arboreal species have first choice, without competition from ground-foraging species (the same individual fruits available to *T. minor* in a tree, if not eaten, eventually fall and become available to the terrestrial species).

I could watch interactions of arboreal *T. minor* in fruit trees without difficulty, but there they had no competition from other treeshrews. Several times I tried watching the ground from discreet lookouts near fig trees (with and without a blind), but both treeshrews and larger mammals tended to be so wary that I collected little useful data. I was more successful at watching interactions at artificial fruit sources (bait stations provisioned with banana for a couple of weeks) from a blind (Poring) or from a distance (Danum). Squirrels and treeshrews came to these baits.

At bait stations that were set up so that only a single animal at a time could feed there (bananas fixed to the end of a single pole), I saw encounters between *T. minor, T. gracilis, T. longipes,* and *T. tana*. Interactions between individuals of different species were all of the simplest and most straightforward kind: the larger species would simply displace the smaller, which would lurk in the background and wait to return when the larger species left. A larger species that arrived when a smaller one was feeding would displace it from the bait by simply approaching. I saw no pursuit of another species that was displaced, and a larger species feeding at the bait appeared to ignore members of other species that hovered a few meters away. The smallest species, *T. minor*, would often suddenly appear at the bait after the departure of another taxon, as if it had been waiting in the wings for its chance. Individuals of all species fed at bait in short bouts of a few minutes, then left and returned later, consistent with the ingestion pattern described above. The behavior at baits, and the fact that radio-collared treeshrews of different species were often together near the same fruit tree, suggests that where fruit is broadcast under a tree there is likely to be little or no direct contest for it among treeshrew species. At a limited source (such as a *Rafflesia* fruit), a larger species would preempt the food from any smaller one.

COMPETITION FOR FRUIT AMONG TREESHREWS AND OTHER TAXA

The fruit species chosen by treeshrews were also favorites of many other mammals and/or birds, and it is unlikely that treeshrews are of more than minor importance as seed dispersers in intact forest. A notable exception to this is for the peculiar fruits of *Rafflesia* spp., which have characteristics that may be best suited to treeshrew dispersal (Emmons, Nais, and Biun 1991). In secondary or highly disturbed forest, where larger mammals and birds have been extirpated by hunting or habitat loss, treeshrews could acquire a more preeminent dispersal role, especially *T. longipes*, which persists well in severely altered forests (chap. 4; see also Stuebing and Gasis 1989). I recorded all mammal species that I saw at fruit sources (Appendix V), and I also usually noted whether birds fed on the fruits. These were incidental notes made while radio-tracking or working on other tasks in the forest, so most are daytime records. Unfortunately, there are no data on the fruits eaten by bats at Danum. The lists of other taxa are incomplete: many larger mammals were shy, did not have home ranges on the study areas, or, at Poring, were absent or rare (e.g., there were no gibbons or orangutans and scarcely any terrestrial large mammals there). I saw other taxa using the following numbers of fruit species also eaten by treeshrews: birds, 11; Prevost's squirrel (*Callosciurus prevosti*), 7; other squirrels, 3; artiodactyla, 3; monkeys, 3; and civets, 3. I suspect that most generalized frugivores (e.g., monkeys, civets, mouse deer) probably eat virtually all of the small, sweet fruits selected by treeshrews. Nonetheless, I think the data give the correct general relationship—that treeshrews show most overlap with birds in fruit species eaten. Despite this, two of the "keystone" species of fruits for treeshrews, *Dialium indum* and *Alangium ebenaceum*, were strictly mammal-exploited fruits, which attracted large numbers of other mammalian taxa, from monkeys to deer (Appendix V), but apparently no birds.

Because treeshrews and squirrels are superficial look-alikes, it is of in-

Diet and Foraging Behavior

terest to compare their diets. Squirrels habituated to us and were easily seen. At Poring we identified nine species of diurnal squirrels (nonflying) on the lower study plot and seven on the upper plot (of which three were different), while at Danum Valley I saw only five species on the study area, but at least three others are recorded elsewhere in the area (Anon. 1993; Appendix III). In this large community of squirrels, only a single species, *Callosciurus prevosti,* fed intensively on figs and the other fruits used by treeshrews, at times evidently all day. This squirrel is a large, strictly arboreal, canopy-feeding species. A small terrestrial squirrel, *Sundasciurus lowii,* the most numerous squirrel in all habitats studied, ate some treeshrew fruits on the ground but did not seem to concentrate much activity around these. In contrast, when acorns were available (Fagaceae, for which treeshrews showed little interest), all but pygmy squirrels displayed frenetic, daylong activity of eating, cutting, and caching them. Treeshrews and squirrels thus had little overlap in frugivory, and the one squirrel that concentrated on the same fruits did so in the canopy vegetation layer, where only one *Tupaia* species feeds.

As well as competing for fruits, in at least one case Prevost's squirrel facilitated feeding for treeshrews. The key species *Alangium ebenaceum* has a small, hard, green fruit. *Callosciurus prevosti* removed and ate the seeds and dropped a rain of discarded fruit fragments to the ground, where the three terrestrial tupaias fed on them. Treeshrews seemed to be attracted to *Alangium* trees where a squirrel was feeding, and I believe that the fruits are too hard for treeshrews to penetrate with their weak jaws, and they may only be able to eat broken fragments. An orangutan chewed and sucked off the edible aril and spat out the seed and pericarp, leaving little of value for other taxa.

In the wild treeshrews and squirrels did not seem to pay any attention to each other, but at an artificial banana bait site with room for one consumer, a horse-tailed squirrel, *Sundasciurus hippurus* (approx. 300 g), completely dominated both an adult *T. tana* and a *T. longipes.* The treeshrews circled in the background, once even growling, and would approach the banana only after the squirrel left. The squirrel did not chase or show any overt gestures of aggression but simply went to the banana (not toward the treeshrew) and the treeshrew left. Even after eating its fill, the squirrel deliberately kept the treeshrews away by dashing in to stand by the banana whenever it saw a treeshrew approach. At another bait site, a *T. longipes* chased off and preempted the banana from a *Sundasciurus lowii* of about half its weight (approx. 100 g), and this squirrel in turn displaced a *T. gracilis*. In this artificial situation, squirrels and treeshrews seemed to behave toward each other as if the taxa were interchangeable, the larger of whatever species displacing the smaller, with squirrels dominant at equal body mass.

The chief indirect rivals of treeshrews for fruit, therefore, seemed to be birds and Prevost's squirrels, followed by monkeys, artiodactyla, and civets. In a direct contest at a limited resource, such as the single fruit of a *Rafflesia* (Emmons, Nais, and Biun 1991), squirrels would probably outcompete treeshrews of the same weight; but this situation is likely to be too rare to have ecological significance except when resources are marginal.

INTERSPECIFIC COMPETITION FOR INVERTEBRATES

In contrast to fruit, for which the consumers can be directly seen for each fruit species, most foragers on invertebrates can be only indirectly assessed by where, when, and how they forage. Two insectivores that take invertebrates from inside rolled leaves at the tips of branches can reasonably be assumed to compete for the same prey. As Root (1967) developed in his classic paper, competition among vertebrates for invertebrates can best be studied within the paradigm of feeding guilds.

We saw above that each treeshrew species has a distinctive tactic for feeding on invertebrates. Each treeshrew that forages in the same microhabitat has a quantitatively and to some extent qualitatively divergent invertebrate diet that should effectively remove it from strong competition with any other: among the three terrestrial species that share the lowland forest floor, one gleans the undergrowth foliage for caterpillars, one gleans ants and termites from substrates other than foliage, and the third digs into the decomposing litter and woody substrate for earthworms and feeds on noxious forms and decapods not taken by the other two. During the study, the only other mammalian species that I saw that foraged somewhat like treeshrews in the same habitat were the diurnal mongooses *Herpestes brachyura* and *H. semitorquatus*, which meandered through the forest, nosing the ground, frequently digging into it for prey. These are likely to compete somewhat with *T. tana* for invertebrates hidden under the surface litter. Unlike large treeshrews, I saw these mongooses enthusiastically digging quite deep holes in the clay below the litter, but I never saw what they pursued.

One tiny squirrel, *Sundasciurus jentinki*, hunts insects in the foliage of small trees, as does *T. minor*, but it only occurs above about 1,000 m

Diet and Foraging Behavior   85

elevation, where *T. minor* is scarce or absent. The rare, shrew-faced squirrel, *Rhinosciurus laticaudatus,* could potentially share the invertebrate resources of *T. longipes* or *T. tana,* but its natural history is almost unknown, and I never saw one. Youngsters of *Sundasciurus hippuru*s did much invertebrate hunting, but they frequented large-diameter understory lianas and treelets, a different substrate and lower level than favored by *T. minor.* I saw one eating a snail gleaned from this substrate, an item that was also found in a *T. minor* scat. *Sundasciurus lowii* and *S. brookei* both forage intensively for an unknown resource on the bark of large trunks near the ground, a substrate/height combination not favored by treeshrews. Lesser gymnures, *Hylomys suillus,* feed on earthworms and a variety of forest floor invertebrates (Payne, Francis, and Phillipps 1985; Tan 1965), but they live above 1,200 m, coincident only with *T. montana* in habitat. Finally, at least five species of true shrews occur on Borneo. Those few that I saw appeared to forage beneath the forest-floor litter, where they could compete with *T. tana.* Their foraging strategies are poorly known, but they appear to feed on miscellaneous arthropods (Tan 1965). I conclude that no other mammals seem to hunt invertebrates in precisely the same fashion, time, and place as treeshrews, although several do so in part.

Birds were the most obvious potential competitors with treeshrews for arthropods. While following treeshrews, I often watched the foraging behavior of common birds, and it was evident that babblers (Timaliidae) often foraged for prey in the same locations as treeshrews do. Like treeshrews, babblers chiefly pick sedentary prey off the foraging substrate and in particular foraged for hidden, immobile prey in sites such as the undersurfaces of curled leaves and interior parts of plants. Bulbuls, in contrast, tended to sally for prey spotted from a perch, while leafbirds and cuckoos took prey from upper, sunlit leaves, places little used by treeshrews, and so forth. The babblers often foraged in mixed flocks of species organized in distinct guilds. Table 5.8 outlines the habits of a few common species of babblers at Danum and the treeshrews with which they are most likely to compete for prey. Birds of this family are extremely common and likely to have a profound ecological impact. A few other bird taxa are also likely to have overlapping foraging roles with particular treeshrews: the pittas (Pittidae) are common terrestrial birds that favor the dank undergrowth microhabitats loved by treeshrews. They feed on invertebrates, especially ants and beetles (Lambert and Woodcock 1996), so they may compete with *T. tana* and *T. gracilis* but especially *T. longipes* for arthropods. Nuthatches (*Sitta*

TABLE 5.8. Some foraging behaviors of selected species of babblers at Danum Valley and the treeshrews they are likely to compete with for prey, as observed during the study by Emmons.

Babbler	Level	Behavior	Treeshrew
Black-throated wren babbler, *Napothera atrigularis*	Ground under litter	Flips over leaves with beak and looks beneath	*T. tana*
Striped wren babbler, *Kenopia striata*	Ground	Lifts up edges of dead leaves, pushes head and beak beneath	*T. tana*
Black-capped babbler, *Pellorneum capistratum*	Ground	Hops along gleaning surface of litter	*T. longipes, T. gracilis, T. tana*
Short-tailed babbler, *Trichasoma malaccense*	Ground		*T. tana, T. longipes*
Horsefeld's babbler, *Trichasoma sepiarium*	0–3 m	Understory stems, leaf undersides, saplings	*T. gracilis*
Chestnut-winged babbler, *Stachyris erythroptera*	Low to midstory	Open forest saplings and leaf bottoms, hanging dead leaves	*T. gracilis, T. minor, P. lowii*
Scimitar babbler, *Pomatorhinus montanus*	Low to midstory	Peers into crevices and holes in trunks and branches	*T. longipes, T. minor, P. lowii*
Gray-headed babbler, *Stachyris poliocephela*	Midstory	Dead-leaf bunches	*T. minor, P. lowii*
Chestnut-rumped babbler, *Stachyris maculata*	Mid-high	Disruptive foliage gleaner	*T. minor*
Fluffy-backed tit babbler, *Macronous ptilosus*	Low to midstory	Dense vine tangles and festoons	*T. minor*
Scaly-crowned babbler, *Malacopteron cinereum*	Midstory	Undersides of terminal leaves and leaf tips	*T. minor*

NOTE: Includes only a partial list of the babblers known from the area (see Anon. 1993, for a complete list).

frontalis) feed on bark invertebrates from tree trunks, where pentails mainly forage, but do it at a different time (day), while piculets (*Saisa abnormis*) search suspended dead twigs, as do both lesser and pentail treeshrews.

A FORAGING ASSOCIATION BETWEEN TREESHREWS AND BIRDS

From the start of the study in 1989, I noted that actively foraging lesser treeshrews were often in the company of mixed-species flocks of birds.

Diet and Foraging Behavior 87

It was only in December 1990 that I realized that particular birds were associating themselves with the treeshrews.

[7 Dec. 90] 1130 h: *T. minor* M91 foraging at 10 m in a dense vertical vine festoon, above a yellow-bellied bulbul; 1200 h: still with birds, 15 m up; 1211 h: still with bird flock, foraging in vines 8 m; 1217 h: M91 seen still with birds 8 m, reaches out head and peers at leaf tips, drongo below, then goes to canopy 10 m (my following broken by heavy rain; M91 went with the bird flock 50–75 m in 3/4 hour). When below, the drongo and bulbul kept looking up at him; 1253 h: still a bird flock around him.

[10 Dec.] 1020 h: *T. minor* M91 insect hunting overhead, a bulbul and a young *Sundasciurus hippurus* also hunting insects near. Goes inside a large dead leaf bunch at 3–4 m. 1044 h: M91 hunting with a bird flock, moving horizontally at 10 m in a dense vine tangle, grey-cheeked bulbuls below. Bulbuls following M91, letting him get ahead, then flying to perch below, etc., etc. They have followed 20 m so far, only bulbuls following M91, other birds go in opposite direction off territory.

[21 Dec.] 0754 h: A *T. minor* hunting vertically in a vine tangle in the subcanopy at 10 m, a yellow-bellied bulbul closely following underneath 1.5–2 m below, hawking insects.

[13 Mar. 1991] 1059 h: Unmarked *T. minor* foraging at 8–10 m, mostly vertically, sniffing in dead, hung up and curled leaves. Yellow-bellied bulbul (banded light blue/dark blue) starts to follow; 1122 h: bulbul still following, *T. minor* at 20 m.

[21 June] 1245 h: Unidentified *T. minor* climbing up and down a gnarled trunk, sniffs in holes. As I watch, a yellow-bellied bulbul flies in underneath.

[20 Apr.] 0811 h: A *T. minor* is foraging at 1 m in a low vine blanket in a treefall gap; a yellow-bellied bulbul is waiting *above,* then flies in under when *T. m.* climbs up high enough for the bird to fit in below.

[13 July] 0914 h: *T. minor* F294 traveling/foraging with a male very close, in a dense vine festoon, foraging in dead leaf bunches. A male racket-tailed drongo briefly below, then a pair of drongos follow *T. m.* to the next tree, and the next, in the vine subcanopy 4–10 m; 0947 h: *T. m.* comes back to same vine tangle, the pair of drongos comes back, still following.

[15 July] 0758 h: An unmarked *T. minor* male foraging at 8 m moving horizontally in a vine tangle, with a pair of RT drongos beneath; a female *T. m.* comes by the same place and the drongos switch to following her instead.

[3 Aug.] 0830 h: *T. minor* M172 in vines 4 m; 0845 h: a mixed bird flock arrives, RT drongos attach to *T. m.* and start to follow, separate from rest of flock.

The field notes above illustrate all of the basic features of the association between birds and treeshrews. Birds of two species that are regularly found in mixed-species flocks, yellow-bellied bulbuls (*Criniger phaeocephalus*) and racket-tailed drongos (*Dicrurus paradiseus*), seek out a foraging *T. minor* and perch about 1.5 to 2 m below it, intently watching

TABLE 5.9. Associations of *Tupaia minor* with birds.

	With YB Bulbuls	With GC Bulbuls	With RT Drongos	With Flock but without Followers
With mixed flock	4	1	4	7
Without flock	12		8	

NOTE: Data from 7 December 1990 to 30 September 1991 and one observation from 1989. Before December 1990 I did not recognize the association with particular species. Total observations of *T. minor* during this period = 140; total noted with any birds = 42 (30% of sightings). Table lists numbers of times obvious bulbuls or drongos were observed following the treeshrew. YB = Yellow-bellied bulbul (*Criniger phaeocephalus*); GC = Gray-cheeked bulbul (*Criniger bres*); RT = Racket-tailed drongo (*Dicrurus paradiseus*).

the treeshrew. The birds allow the treeshrew to advance ahead, then fly to the next perch below. The rummaging tupai flushes hidden insects, which the birds sally to catch. Many insects have an escape behavior of dropping down when disturbed from foliage (pers. obs.), and these two (follower) bird species profit from the disruptive foraging of *T. minor* by simply waiting, flying from one perch to the next, keeping below the mammal.

This enterprise must be a successful strategy for the birds, because when locked onto a treeshrew, they abandon their mixed flock to follow it. The pattern of observations (table 5.9) shows that both species of follower birds hunted under lesser treeshrews alone two or three times as often as when they were with mixed bird flocks and that, conversely, lesser treeshrews, when accompanied by birds, more often than not had evident followers hunting below them. On all occasions except that described above for 7 December, birds of only one species at a time followed a treeshrew. Once I saw a yellow-bellied bulbul foraging below a chestnut-winged babbler (*Stachyris erythroptera*) with the same behavior exhibited when foraging below a treeshrew.

During the seven months when the behavior of followers was recorded, there was a seasonal shift in the bird species that accompanied lesser treeshrews (fig. 5.9). I saw no drongos follow in April, and no bulbuls in May, July, or August. There seem to be no published reports of breeding cycles for these birds in Borneo, but in a large series of females collected in Sabah by F. Sheldon (pers. com.) during all months, the only reproductive female yellow-bellied bulbul (which had just laid an egg) was collected in July and the one racket-tailed drongo was collected in April. This hints that each species may have stopped following when it was busy nesting.

Apart from the clear foraging link of the follower birds with *T. minor*, I was not able to identify the basis for the general associations be-

Diet and Foraging Behavior 89

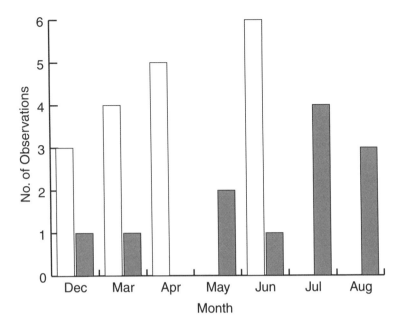

Fig. 5.9. Number of times that bulbuls (open bars) and drongos (shaded bars) were seen following *T. minor* at Danum Valley in months of 1991. The amount of observation time per month was not equal; the data show the relative monthly contributions of the two bird taxa.

tween lesser treeshrews and whole mixed flocks. Treeshrew and flock sometimes moved together for long periods and distances (up to 1.5 h and more than 100 m). I do not know if the other birds also escorted the treeshrews, or pursued the follower species, or if the treeshrew may sometimes even follow the birds. I never saw the treeshrews paying any obvious attention to the birds, although they would have been well aware of them. Squirrels in Sabah also can associate with mixed bird flocks. *Sundasciurus jentincki* usually seems to hunt insects by foliage gleaning in the crowns of treelets from within bird flocks (pers. obs.; Payne, Francis, and Phillipps 1985); and young *Sundasciurus hippurus* also sometimes insect forage with an escort of birds (older *S. hippurus* seem to spend less time hunting invertebrates).

APPENDIX: METHODS

The feces in the trap were collected each time a treeshrew was captured (large clean leaves were placed under the traps to catch them), but not all captures produced samples. Because of the rapid food transit time (Emmons 1991), both feces containing prior meals and those containing only bait (banana) were usually together. Only feces that included something other than bait were preserved, by placing them in Whirl-Pak™ bags, in ethanol, for a total of 185 samples. Warren Steiner identified and counted invertebrate fragments to the nearest practical level (order, family, subfamily). Ingested earthworms (Oligochaeta) were easily recognized by the presence of their distinctive, sigmoid setae, although nothing else identifiable remained of them. For reference, earthworms collected at Danum were completely digested in NaOH.

As a cautionary note for other field biologists, I reemphasize that it would not have been possible to predict the differentiating features of treeshrew diets or foraging modes either from their morphology or from field observations. Only the tedious labor of microscopic examination of fecal samples yielded the key information about what the species actually ate.

CHAPTER 6

Nesting Behavior

Because half of a treeshrew's time is spent inactive, its nest site and the nest itself are primary defenses against exposure and predation. The absentee maternal care system of treeshrews is linked to their nesting behavior as young are placed in one nest and their mother uses another (Martin 1968). To study the maternal care system, I therefore needed to try to find nests of both adults and young, and learning where treeshrews nested was a pivotal element of the field study. An obvious question about the absentee system is whether treeshrew nests are placed where predation risks can be expected to be especially low (or high). The nesting behavior of treeshrews in the wild was known only from the handful of scattered notes that are collected in the following few passages:

> [A pentail treeshrew was] caught in a nest made of leaves and fiber in a tunnel 2 feet long in a hollow branch of a tree. (Thomas 1910: 426)

> In 1954, the aborigine trapper, Inche Sipang anak Ecoin caught four specimens of Pentail tree-shrews in a single nest hole of a tree trunk which he cut down. Two nests examined consisted of dried leaves, twigs and fibers of soft wood. (Lim 1967: 376)

> Cantor writes on the habits of *Tupaia [glis ferruginea]*. . . . [N]ight sends them to sleep in their rudely constructed lairs in the highest branches of the trees. . . . Of *Tupaia glis ferruginea*, Robinson and Kloss remark ". . . The nest is found in holes, often in fallen timber." (Cited in Lyon 1913: 21)

> [A Philippine treeshrew, *Urogale everetti*,] lived for several days nearby before he was seen. His hideout was found to be in the bottom of a pit, the sides

of which contained numerous holes, presumably of rodent origin. Natives state that tree shrews nest in the ground or in cliffs. (Wharton 1950: 352)

[*Tupaia minor*] breeds in a nest in an old stump covered in creepers, but I am not sure whether it makes the nest itself or occupies the nest of a bird. I have found two of these nests, but the material used was different. (Hose 1893: 30)

[D'Souza (1972) found three nest sites of *T. minor*. Two were in holes in living trees, at 19 and 20 m, the other was in a crevice of dried palm leaf at 13 m, an] isolated tree surrounded by thick undergrowth. [The nest was] constructed of tightly woven dried grass, leaves, and strands from the nylon rope supporting the transect. . . . Although the female inhabitant of this nest was seen taking nesting material to this nest, it could be that originally this "woven" nest was inhabited by some other mammal or bird species. Laboratory data show that *T. minor* consistently builds a nest of loosely packed leaves and other fibrous material, but has never been noted "weaving" the materials. (Pp. 178, 182)

We therefore had some idea of where we might look for nests for only two of our study species. From the first day that the first treeshrew was fitted with a transmitter, we launched into intensive action to find the nest sites of every treeshrew during the first field season (see chapter appendix). Our simple strategy was to home in on the radio signal after the treeshrew had retired to its nest.

NESTING SITES

PTILOCERCUS LOWII We found only a single sleeping site of pentails, at Danum Valley. This was inside a giant, emergent, living tree most of whose main trunk was apparently hollow. All four captured pentails used this same den. During the day their radio signals showed them to be extremely high in the tree, perhaps 30 to 40 m or higher. When they became active, they descended and exited onto the outside of the trunk either at a high, undetermined spot above 10 m or at one of two small holes in the buttressed base of the trunk, where we saw them emerge (fig. 6.1). I could not tell whether the several individuals that shared this tree slept by day together in the same spot, or just in the same general region of the tree, but their radio signals seemed to come from the same place.

TUPAIA MINOR Of the eleven nest sites found for four lesser treeshrews, eight were high (above 15 m) in large, living emergent trees and could not be pinpointed. Several of these trees had visible holes in the trunks, or were wrapped by vines or the roots of strangler figs that created many crevices where a nest could be hidden. One sleeping site was

Nesting Behavior

Fig. 6.1. The base of the pentail nest tree. One of the exit holes is visible in the lower center, and a trap is attached to the sapling and liana at left.

at about 10 m in the vine-covered top of a rotten dead tree that was hung up on its neighbors leaning over a ravine; one was low, at 3 m in a dense vine festoon on a small dead tree in a canopy vine blanket; and the last was on a huge, vine-wrapped stump. I did not dismantle any of the sites to investigate the exact sleeping place.

TUPAIA GRACILIS   Ten of the eleven nest sites discovered for four slender treeshrews were in the undergrowth near the ground at 1 to 5 m height ($\bar{X}$ = 3.2 m, SD 1.4). The exception was high in a horizontal branch of a large emergent tree. The ten understory nests were in the following sites: two in hollows within slender, vertical, standing dead stumps; five exposed in the tops of small saplings; three in vine festoons in the undergrowth or on top of understory treelets; and one in a hole in a vertical trunk of a living tree. For eight sleeping sites the spot used was identified, and in seven of these there was some type of nest. These nests were constructed in several ways. The most often used den of F105 was in a slender, 15.5 cm diameter stump, at a height of 4.8 m. The nest was a 10 × 10 cm ball of teased fibers, with no leaves, wedged tightly into and filling the hollow. This nest differed from four others examined, which were wholly of leaves, with no teased fibers. One nest used by Mfoot was hidden at 1 m in a

TABLE 6.1. Dimensions (cm) of six excavated burrows and included nests of *T. longipes* at Poring.

Entrance	Tunnel Length[a]	Chamber Size	Chamber Depth	Nest
6	60, 50	23 × 16	35–40	15 × 22
4.5	60, 24	46 × 11	40	15 × 11
5	30	30	30	none
4.5	40, 47	50 × 40	30	15 × 14
5	70, 30			none
5, 4.5	140, 118	15	35	15 × 20

NOTE: The third and fifth, among roots under trees, could not be completely dug out. Chamber depth is measured from the ground surface to the top of the nest chamber.

[a] Measurements of the two arms of the tunnel, from the nest chamber to the entrance hole.

pile of leaf debris in a thicket of vine stems and was a constructed ball of leaves, 12.5 cm in diameter, with a nest chamber of 4.5 cm diameter. One leaf nest was a tiny, rudimentary cup or ball (which disintegrated when examined), and one seemed to be unconstructed: "No nest, a bunch of large, heart-shaped vine leaves stuck together with mycelia, to form a tiny shelter only 10 cm across, with two large dead leaves covering a platform of a few leaves that look sat upon, and smell slightly of *Tupaia*." This platform of leaves was caught in a tree fork at 2.5 m. Two constructed leaf nests were not disturbed, so their interior structure was unknown, and one sleeping site was in a woven, hanging bird's nest.

Slender treeshrews thus usually nested in the understory but above the ground, either in hollows in standing living or dead trees or stumps of small diameter or in exposed leaf nests in the tops of treelets or in vine tangles. Several of the nests were in stumps or treelets isolated from surrounding vegetation. This species alone had a tendency to jump from its nest and land on the ground with a loud thump, emitting alarm calls, if I approached its nest at night. This happened four times with collared individuals and I believe two or three other times with unmarked ones. It was thus difficult to penetrate the undergrowth and precisely locate nests (the animal would jump out before I could get close enough to find it). Few nests were found, as I was reluctant to disturb or expose treeshrews at night.

TUPAIA LONGIPES Thirteen sleeping sites of plain treeshrews were in underground burrows, and one was in a hollow log 50 cm above ground. The latter was used but once, when the animal had been disturbed from near its burrows at nightfall. All other sites identified, on all nights for

Fig. 6.2. Alim Biun pointing to the entrance of a *T. longipes* burrow, which is just beyond the tip of his finger; leaves have been slightly cleared away. The entrance is virtually invisible.

which dens were known, were underground. Six burrows were excavated at Poring, and the other seven were inspected from the surface.

Tupaia longipes burrows were of the simplest type: each was a plain tube with an entrance at each end and an enlarged nest chamber somewhere toward the middle. The tube varied from a straight line with a slightly enlarged area in the middle (the last nest; table 6.1) to a U- or Y-shape, with two tunnels angling down from the chamber. Some of the burrows had chambers under trees, stumps, logs, or roots, but often they

Fig. 6.3. A nest constructed only of overlapping leaves from a burrow of *T. longipes*.

were in the flat forest floor among ordinary understory saplings. The burrow entrances were the most well hidden of any mammal burrows of my experience. They usually opened into flat ground of a slope and were hidden in the leaf litter, unmarked by any feature (fig. 6.2). Although we knew the exact spot above the nest chamber where the collared treeshrew had slept, it sometimes took 20 min to find an entrance. The second entrance was usually plugged with leaves and even more inconspicuous than the used one. Without radiotelemetry it would have been impossible to even suspect the existence of these burrows, let alone find them. By frequently switching burrows, plain treeshrews perhaps avoided creating visible runways at the entrances.

Four of the six burrows contained leaf nests (fig. 6.3), some in good condition, others of just a few, decomposed leaves. The best-constructed nest (the first listed in table 6.1) was made of about eighty large (11 × 23 cm) dead leaves that were simply overlapped and filled the chamber. The roof was of newer leaves, with older, broken leaves on the chamber floor below intact leaves. The leaves included *Aprusa* sp., *Baccaurea* sp., *Shorea argentifolia,* and others from shrubs, trees, and lianas in the immediate neighborhood, so that the treeshrew could have picked up the fallen leaves from close to the burrow. None of the nests had any kind of lining material. Many of the nests and nest chambers were damp at

Nesting Behavior

Fig. 6.4. Nest site of a *T. montana* in a ground cavity at the base of a tree. The entrance is above the tip of the pale tape measure at left center.

the time of excavation in the wet season (weeks after their recorded use in the dry season).

TUPAIA MONTANA   We found nine sleeping sites used by five mountain treeshrews. All were leaf nests. Seven of these were hidden or partly hidden at the ground surface underneath the surface-layer organic mat of moss, rootlets, and leaf litter that is characteristic of montane forest. None of the nests was in any kind of excavated burrow in the soil, but they were placed in natural crevices or cavities under the moss mat: between rocks, under roots, or pushed under moss (fig. 6.4). The other two nests were inside the hollow of a fallen log on the ground and inside a hollow tree stump. Some nests were completely hidden and undetectable visually, with completely invisible entrances, while others were so close to the surface, or in such small crevices, that part of the leaf nest stuck out into the open.

The six mountain treeshrew nests that were excavated were of different construction: two had quite thin outer layers of large overlapping leaves that surrounded large balls of fibers; two were spheres of packed, overlapping leaves but had no interior lining; and two were balls of fibers with no, or few, outer leaves (fig. 6.5). These last were placed in unusually protected, dry sites inside a stump and under large roots. I describe one of each type as examples. (1) A nest of Tm F150 was on a steep hill-

Fig. 6.5. A nest of *T. montana,* constructed only of woody fibers.

side, on the ground surface on rocks at the foot of a sapling, with a vine, roots, and natural leaf litter covering the top (fig. 6.4). It measured 20 × 15 cm and had a small entrance chamber leading into a 10 × 8 cm nest chamber in the back. The nest was of packed, overlapping leaves of many types, including rattan, bamboo, ferns, and ordinary large tree leaves. (2) A nest of Tm F154 was underground, 5 cm beneath the surface on a moderate, rocky, rooty slope. The outside entry to the cavity was 7 × 7 cm, and the nest was 7 cm from it, wedged into a large hollow that it did not fill. The 18 × 13 cm nest had a superficial layer of large leaves, mainly on top, surrounding a large ball of fibers. (3) A nest of Tm F154 was under large, ramifying tree roots. I could not reach it, but the nest was a round sphere of about 13 cm that was lying superficially on the ground within a large cavity. It appeared to be of teased wood fibers.

In this small sample there was no evident correlation between nest type and sex of the treeshrew. Males used a completely fiber nest, and one with only leaves and no fibers, whereas females used the other four, of all types.

TUPAIA TANA We discovered twenty-three sleeping sites of large treeshrews, and the nest itself was described for eleven of these. Of the twenty-three sites, seven were in standing dead stumps; five were in fallen dead logs; five were in hollows of living trees (fig. 6.6); three were exposed in the tops of saplings (fig. 6.7); two were in roots of fallen staghorn

Fig. 6.6. The tree hole that contained the nest with young of *T. tana* F109.

Fig. 6.7. An exposed leaf nest of *T. tana* F109, in the tree fork above Alim Biun, at the upper edge of the photo.

ferns; and one was in a treefall tangle. Only four were exposed; the other nineteen were sheltered in a hollow or cavity of a live or dead tree. The height of nineteen nests ranged from 0.2 to 8 m ($\overline{X}$ = 3.4 m, SD = 2.5), but the two highest nests are omitted because their heights could not be estimated from the ground (see chapter appendix). All usual sleeping sites were above the ground, but the mother F109 spent one night on the ground with her young as a special case (Emmons and Biun 1991; see also chap. 10). The standing stumps and den logs were often rotten snags of large emergents, honeycombed with cavities. Nests were usually wedged into a tight cavity, or in one case under a large flake of bark.

The three leaf nests exposed in the crowns of saplings were flattened balls of large outer leaves surrounding an inner lining of fibers. In one of these, an old, decomposing nest, there were many twigs incorporated into the outer leaf ball. This was unique among all treeshrew nests dissected. Three of four nests that were inside cavities were balls of teased fibers with only a few leaves on the outside. One nest in a cavity had been recently rebuilt and included both green and dry leaves of bamboo and other plants on the outer shell. The fiber liners were made with coarse, short strips of wood or bark, perhaps picked from the interior of the decomposing logs that housed them. The mean outer dimensions of four nests were 27 × 17 cm; the fiber liner of one was 6 cm thick, and nest chambers measured about 11 cm in diameter.

USE OF SLEEPING SITES

The pentails all used the same sleeping site on every day recorded, except on a single day (8 Dec. 90) when we arrived at 1836 h to find F96 already active about 200 m from the nest tree, suggesting that she had spent that day near this far place. Because her den site for that day is unknown, it is not recorded in table 6.2.

All individuals of every *Tupaia* species followed had multiple sleeping sites and moved from site to site almost daily (table 6.2). The longer an animal was tracked, the more sites it used, so based on the small sample sizes for individuals, it is impossible to estimate how many total sites any treeshrew may have had. *T. gracilis* F67 used the most sites—14 during 16 recorded nights—while *T. tana* F109 used the fewest—6 sites during 24 nights. Although in 1989 at Poring I suspected that our behavior of always searching for exact sleeping sites at night (albeit always as silently as we could) might be causing treeshrews to change

TABLE 6.2. Numbers of sleeping sites (SS) used by treeshrews during radio-tracking.

Species	No. of Individuals	Nights Recorded	SS Used	Nights/SS
P. lowii	4	44	1	44
T. minor	5	38	23	1.7
Females	3	27	17	1.6
Males	2	11	6	1.8
T. gracilis	4	44	33	1.3
Females	3	37	29	1.3
Males	1	7	4	1.8
T. montana	5	36	24	1.5
Females	3	21	16	1.9
Males	2	15	8	1.3
T. longipes	8	66	43	1.6
Females	5	47	32	1.4
Males	3	19	11	1.7
T. tana	19	128	75	1.7
Females	11	81	42	1.9
Males	8	47	33	1.4

NOTE: Largely from periods when animals were continuously followed. All pentails used the same site on every recorded day (see text).

sites, data from Danum, where I did not search for nests, show equivalent turnover of sleeping sites even when animals were undisturbed. For example, at Poring (1989) all *T. tana* shifted sites at a rate of 2.1 nights per site, whereas at Danum (1990–91) they moved at 1.5 nights per site or more often. Likewise, *T. longipes* at Poring averaged 1.4 nights per site and at Danum 1.6. Males and females of a species behaved alike within the samples (but males were tracked on fewer nights). Sleeping sites of *Tupaia* were scattered all over their home ranges, with no particular pattern.

Except for a single pair of *T. longipes,* in which the male and female used three of the same burrows (but never at the same time; see chap. 8), none of the 204 sleeping sites used by *Tupaia* species was ever recorded to be used by more than a single individual. However, sleeping sites of treeshrews of the same or different species could be so close together, such as in the same treefall tangle, that they could not be distinguished by radiotelemetry without approaching to within a few meters.

NEST BUILDING

I never saw a treeshrew building its nest, but when *T. tana* F109 visited the nest cavity to nurse her young, she at least once carried a single leaf in her mouth and left the cavity without it (videotaped; Emmons and Biun 1991; see chap. 10 below). At the end of a day a *Tupaia* would almost always run straight to its sleeping site and immediately cease moving, but in rare instances a *T. tana*, *T. gracilis*, or *T. longipes* continued activity at its den site for half an hour or so, running back and forth or up and down around its nest. I interpreted this as probable nest-building activity but cannot verify it.

DISCUSSION OF NESTING BEHAVIOR

SLEEPING SITES

For the two arboreal species there were accurate descriptions in the literature of both sleeping sites and nests. Fortunately, those accounts describe nests of species for which I have the least data, so together, both sleeping sites and nest structures are known for all six study species.

All six species had distinctive kinds of sleeping sites. These fall into three sets of pairs with similar habits (table 6.3). The preference of the two arboreal species for nesting in canopy tree hollows, as do most other canopy small mammals, may simply indicate that tree hollows are the safest, warmest, and driest places to nest in the canopy. But pentails have thus far been reported to nest only in tree or branch hollows, while *T. minor* will use other types of sites, some of them, like those of other *Tupaia* species, exposed in the midstory or understory foliage.

Slender treeshrews more often than any other species used exposed sleeping sites in vine tangles or treelets, and they seemed to use opportunistic shelters such as a bird's nest or a pile of leaf debris. The frail and unprotected nature of these sites is, I believe, the reason that if disturbed, this species alone customarily abandoned its nest with a flying leap into the night (it was clear from what I could hear that the animals could not see where they were headed in the dark). Large treeshrews preferred an array and height of sleeping sites similar to those of slender treeshrews (see table 6.3), but most were protected within woody crevices or tree hollows. I crashed around many of these at night, without any of the animals moving, but once, inadvertently, I brushed by a *T. tana* that was resting in an exposed leaf nest in a low sapling in late afternoon, and it jumped to the ground whistling.

TABLE 6.3. Summary of nest site locations and nest constructions.

Species	Nest Location	Nest Structure	
		Leaves	Fiber Lining
P. lowii	High canopy tree hollows	+	+[a]
T. minor	High canopy tree hollows, vine-covered understory stumps or treelets	+[b]	
T. gracilis	Understory exposed in treelets or vine tangles, hollows in vertical stumps or trees, opportunistic shelters	+	+ or -
T. tana	Understory to midstory in hollow logs and stumps or hollow trees, rarely exposed in treelets	+	+ or -
T. longipes	Underground in dug burrows	+	-
T. montana	Under the organic mat on the ground surface, in stumps	+	+ or -

[a] From Lim 1967.
[b] From D'Souza 1972.

It is a curious circumstance that *T. tana*, which has elongated claws, digs to forage, and is the most terrestrial species, nests only above ground; whereas *T. longipes*, which does not have large claws, does not dig to forage, and is slightly less terrestrial, lives in underground burrows. Morphology is thus not a good predictor of burrowing behavior. Use of below-surface nests by *T. longipes* and *T. montana* might seem to be evidence for an evolutionary connection between these two, but this similarity is only superficial, because the habit of building a nest in a crevice is common to several *Tupaia*, but actual burrowing is thus far known only in *T. longipes*. *Urogale everetti*, likewise said to live in holes in the ground (Wharton 1950), was thought by Lyon (1913) (with reservations), to most closely resemble *T. tana* (with which it shares elongated foreclaws). However, DNA hybridization studies suggest that the Philippine genus *Urogale* and the *Tupaia glis/longipes* species group form a clade together distinct from that of *T. tana/montana* (Han et al. pers. com.). If so, underground burrow nesting may be a phylogenetic trait.

Our finding of a group of pentails that shared a sleeping tree conforms to the reports by Lim (1967), Muul and Lim (1971), and Gould (1978) that this species nests communally. Because pentails have been captured in groups in a nest and share nests by choice in captivity (Gould 1978), I assume that the pentails we found at the same den site were nest-mates.

Nesting Behavior

The nesting behavior of pentails is thus divergent from that of *Tupaia* species in that no *Tupaia* adult was ever recorded sharing a nest with another individual, and whereas pentails used only one sleeping site, *Tupaia* each had many sites and changed them often.

NEST STRUCTURE

Except in a couple of cases in which treeshrews were clearly squatters in old bird nests, all of the nests found in the wild for all treeshrew species were quite similar: a sphere with an outer shell of large, simply overlapped leaves surrounding a ball of woody fibers. The sleeping chamber was always very small, a close fit for a single individual. Both layers together, or either layer alone, could be present. However, many den sites had either old, decomposing, flattened bits of nests or no nest material at all (in some cavities). Treeshrews are not compulsive housekeepers.

The combined data from all species show that the kind of nest used is associated with its site. Exposed nests always had thick outer layers of leaves and also fiber liners. Nests in dry tree cavities usually were balls of fibers with few to no outer leaves. Subsurface nests always had outer leaves but only sometimes fiber liners. Outer leaves thus seem to be incorporated into nests exposed to rain or wet ground but are omitted from some dry sites. Only *T. longipes* was never found in nests lined with fibers, but the sample size is too small to conclude that the species does not use such nests.

From the earliest reports, naturalists repeatedly speculated that the nests used by treeshrews are not of their own design and workmanship but those of other species (Hose 1893; Wharton 1950; Lim 1967; D'-Souza 1972). Captive *T. glis* studied by Martin (1968), like other treeshrew species in captivity, carried real or fake (paper) leaves into their boxes to line their nests. Leaves were carried one by one in the mouth, as I saw *T. tana* F109 do in the wild. In a frenzy of construction, a prepartum female might carry as many as four hundred pseudo-leaves per day into her sleeping site (Martin 1968). I believe that there can be no further doubt that the simple nests of loosely overlapped leaves, such as those used by *T. longipes* and often *T. montana* in Sabah, are made by the treeshrews. The question remains, Do treeshrews also make more complex nests including teased or picked fibers? And could *T. longipes* dig its own burrows with its unlikely morphology?

If the nests that treeshrews use are appropriated from another taxon, then they must be squirrel nests, for these have the same construction of a ball of leaves lined with fibers and are of the same general size and, often, placement (Emmons 1975). Malaysian squirrels usually make exposed leaf nests by clipping whole leafy branch tips and constructing the outside with these, so that the outer shell incorporates many twigs (Payne 1979), but the nests of a number of species of Bornean squirrels have apparently not been described. Because some African squirrels make nests of individual leaves picked up from the ground (Emmons 1975), it cannot be concluded that Bornean squirrels do not make some nests without twigs. The one *T. tana* nest that included many twigs I presume likely to have been made originally by a squirrel. The linings of squirrel nests are often made from teased strips of bark or fiber (Payne 1979; pers. obs.) that might be longer than the short, coarse fibers of treeshrew nests, but again we lack information on the nests or nest sites of sympatric squirrels.

The burrows of *T. longipes* were all of the same size, design, and placement, and I believe it likely that they were dug by the treeshrews themselves. Otherwise one must conclude that plain treeshrews borrowed only burrows from one rodent genus. No rat nests, seeds, or other evidence of rats was present in any burrow, and in general murid burrows tend to be more complex. Because the burrows of Bornean rodents are undescribed, we cannot say whether the dens of plain treeshrews resemble those of another species. No Sabah squirrel is known to burrow, but no radio-tracking studies that would discover this have been done. The fact remains that no treeshrew was ever seen digging into the mineral soil of the ground.

The distinct and characteristic sleeping sites chosen by each treeshrew species, coupled with the fairly uniform size and type of nest used, is in my view evidence that treeshrews mostly choose sites and make their own nests in them. Otherwise, one would be forced to assume that each treeshrew species has adopted the nests/nest sites of a particular squirrel species or set of species of about its own size, an unlikely scenario. Nevertheless, each individual *Tupaia* has a large number of sleeping sites, and some of these may be opportunistically appropriated from other taxa while most are furnished by the treeshrews themselves.

Nests of the common treeshrew (*T. glis*) are apparently undescribed apart from the passages quoted at the beginning of this chapter. No hint of their nesting in burrows is to be found, although the "holes" of Robinson and Kloss could refer to burrows. It is thus not possible to say whether *T. glis* and *T. longipes* have different nesting habits. Without radio-tracking, burrows would never have been suspected for the treeshrews in our study, so a lack of evidence is inconclusive.

In a remarkably parallel pattern to that of these treeshrews, I found that a ground-foraging tree squirrel in Africa, *Funisciurus pyrrhopus,* a species that feeds largely on termites, digs simple burrows just like those of *T. longipes,* while its three sympatric congeners sleep in exposed, aboveground leaf nests as do many other squirrel species worldwide (Emmons 1975). All *Funiscurius* spp. were scansorial and foraged much on the ground, but burrowing by *F. pyrrhopus* is certainly a derived character. It is an intriguing convergence that *T. longipes,* the only treeshrew that we found to eat many termites, has also taken to living in burrows, most likely also as a derived behavior. Perhaps digging into termite mounds for prey leads easily to a burrow-nesting lifestyle.

NEST SAFETY

The nest sites of *Tupaia* in burrows, tree hollows, and leaf nests in trees, vines, and treelets do not exhibit any obvious safety features or special elements compared to the nest sites of other small mammals. More exposed sites, such as those used by slender treeshrews and mountain treeshrews, are unprotected from almost any significant predator (mammals, snakes, ants), while those hidden within hollows are somewhat more secure from certain threats but by no means safe. Many nest hollows or crevices were in easily pulled apart rotten stumps and did not have tight entrances. The varied siting of *Tupaia* nests thus does not yield many clues about the function of the absentee maternal care system, except the fact that the nests are not unusually safe.

Ptilocercus nests may be much better protected. If pentails nest inside small, high canopy hollows reached only from the inside of hollow trees, their nests should be safe from attack by nearly all vertebrates.

THE EVOLUTION OF SMALL MAMMAL NESTING

Small homeothermic mammals almost universally build and use nests or sleep in dens both to conserve their hard-won calories and escape predation. The large relative body surface areas of small mammals require behavioral or physiological steps to prevent heat loss during periods of inactivity. The most primitive living small mammals all build nests, including marsupials (Emmons and Feer 1997), shrews (Churchfield 1990), elephant shrews (Rathbun 1979), and tenrecs (Eisenberg and Gould 1970). Virtually all of the small rodents build nests, and even some large mammals, including pigs, build nests to insulate and shelter their altricial young. Bats, with a morphology nonconducive to nest construction, huddle, group inside cavities or caves, or allow their body temperatures to drop, and fall into torpor to conserve energy. Bat and marsupial altricial young are warmed by attaching to the maternal body, conveniently, by their feeding tubes. That treeshrews use nests, and put their altricial young into them, is therefore the expected mammalian situation. It would be strange if they did not themselves build the nests they use, as do virtually all nesting mammals. Does this have any bearing on evolutionary trajectories that could lead to primatelike lifestyles?

A few of the smallest primates, including lemurs (e.g., *Cheirogaleus* spp., *Microcebus* spp.) and galagos (*Galagoides* spp.), and some larger species such as aye-ayes (*Daubentonia*) make leaf nests for themselves and for their newborn young (Charles-Dominique and Bearder 1979; Doyle and Martin 1979; Wright and Martin 1995). However, both galago and tarsier mothers do not leave their older young in the nest while they forage at night but carry them out and "park" them on an isolated branch (Charles-Dominique and Bearder 1979). This implies that it is somehow safer for the young to be outside the nest when the mother is not there, despite the loss of warmth. Galagos and some other prosimians carry their small young in their mouths, like nonprimate mammals, as do a number of lemur species (Klopfer and Boskoff 1979), but newborns of many species, and all older infants, cling by their hands and feet to their mother's fur. A young mammal attached to the maternal body is probably safer than in any other place. The development of strong grasping digits and a larger adult body size liberates all diurnal primates and most prosimians both from the need for a nest to house their young and from having to leave them alone, exposed to predators.

The living primates thus exhibit the evolutionary progression from small-bodied nesting mothers that separate from their newborns to forage, as do most other small mammals, to larger species with attached, clinging young, emancipated from nests. Treeshrews are not only far off from this trajectory of increasing maternal physical contact/attachment by the young, but they appear to have diverged in the exact opposite direction, since they have lost the behavior of ever carrying or retrieving the young and do not even share their own nests with them (Martin 1968).

APPENDIX: METHODS

The effort to describe nest sites was concentrated in the first field season (1989), at Poring. We found nest sites by homing in on the signal of radio-tagged *Tupaia* at night after they had ceased activity and pinpointing the exact location of the transmitter. When a nest is not in plain sight, finding it entails circling the spot to bring the receiving antenna tip close to the hidden transmitter. Finding the exact sleeping place by telemetry alone is possible only if the animal is near the ground, because the precise site of a transmitter high in a tree cannot be identified with a receiver from below. Sleeping sites were marked with flagging, and at the end of the 1989 field season, after we had finished radiotelemetry of the animals, some sites were dismantled or excavated to find the exact sleeping spot. This destructive sampling was done only at Poring. Because treeshrews often changed a sleeping site on the night after we had tramped around to find it, in 1990–91 at Danum Valley I did not seek out many nest sites, so as to avoid influencing movement patterns. Thus only a few nests were found at Danum. As all of these were situated like those of the same species at Poring, all sites from both areas are combined in the descriptions.

CHAPTER 7

Activity Patterns

Even without other knowledge, a description of the activity pattern of an animal reveals much about its life. As in human endeavors, the amount of time animals spend in various activities is a direct reflection of their basic economics. The activity of treeshrews was recorded by following radio-tagged individuals, rain or shine, from before they left the nest to begin their daily activity until they returned to the nest at its end. The results below are compiled from 185 complete daily records (see chapter appendix). Because it was rare to see the animals, the "activity" described here is an analysis of the movements of radio transmitters through the forest.

LENGTH OF THE ACTIVE PERIOD

All *Ptilocercus* individuals were active only at night, and all *Tupaia* individuals were active only by day. The time that activity began was almost fixed, but the time that it ended was highly variable.

Pentails left their nest high inside their giant hollow home tree and came down the tree after complete darkness had fallen, at a mean time of 1840 h (18.67 ± 0.31, N = 26; range: 18.27–19.17 h). There seemed to be a shift in exit time corresponding to the small shift in sunset during the year: activity began at a mean time of 1825 h in November and December (N = 7); at 1858 h from March through June (N = 9); and at 1833 h in August and September (N = 8). Because a different pentail was

Activity Patterns

tracked during each period, these differences could be individual rather than seasonal, but as several other pentails often appeared outside of the tree at the same time as the radio-tagged individual, I believe they exited synchronously.

All individuals of all *Tupaia* species left their nests just as the first gray light touched the forest floor, most at 0550 to 0555 h (the canopy lightens a few minutes earlier than does the ground). Exits were tightly clustered between 0545 and 0605 h. Adjustments to the length of the active period occurred through variation of times of return to the nest (fig. 7.1), never by changing the time that activity was initiated.

Pentails had the tightest grouping of times of return to the nest (fig. 7.1), with half of all records in the 15 minutes from 0515 to 0529 h and three-fourths of records within the last half hour of darkness, 0500 to 0530 h. When pentails returned to their sleeping site, the sky was already pale from the breaking dawn and there was far more light than there was when they emerged at nightfall, when to me it seemed pitch dark.

The diurnal *Tupaia* species were more variable in their behavior. Because emergence times were virtually constant, differences in the total length of the active period largely reflect differences in times of return to the nest. Females all had modal times of return to the nest within 15 minutes of 1800 h, but median times were earlier, reflecting the two- to three-hour spread in time when animals retired. For most species, there are too few records to distinguish the behavior of males and females, but for *T. tana*, males usually entered the nest 1.25 hours earlier than females. (See table 7.1, fig. 7.1.) When total length of the daily active periods are compared within species, those of males are significantly different from those of females only for *T. tana* (Mann-Whitney U test $P < 0.00$). In between-species comparisons, of females only, *Ptilocercus* is significantly different from all other species and *T. tana* from all but *T. longipes* (Mann-Whitney U test $P < 0.05$). These latter two are active for fewer hours per day than the others.

RESTING

During the day, *Tupaia* species rested from time to time. Those that I saw (*T. tana* and *T. minor*) crouched on a branch or log, with the tail curled snugly forward around the feet. They rested with eyes wide open, completely immobile and blending invisibly into the background. Lesser treeshrews that were in fruit trees would rest in a spot sheltered by leaves or branches from above and periodically get up, scratch and groom, eat a

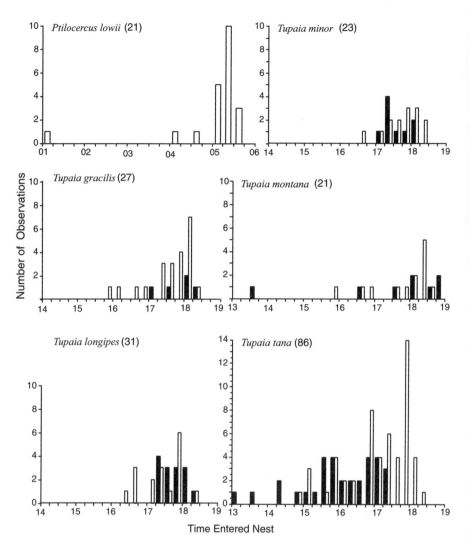

Fig. 7.1. Times at which radio-collared treeshrews entered their nests at the end of activity, grouped by 15 min intervals. Females = open bars; males = shaded bars.

fruit or two, and then return to the same perch for another interlude. A lesser treeshrew would use the same resting perch when feeding in a fruit tree on successive days.

Tupaia species were never known to rest by day in a nest, although a few times *T. minor* rested in a tree where there was a sleeping site and thus could have been in a nest (it was impossible to precisely pinpoint

TABLE 7.1. Activity statistics of treeshrews.

Species	Total Active Period (h)	SD	No. of Days	Time Resting (h)	Median No. of Rests	No. of Days	End Activity, Median (h)[a]
P. lowii, all	10.29	1.09	20	1.61	1	20	0515
T. minor, all	11.81	0.56	20	1.25	1.5	20	1730
Females	11.88	0.61	13	1.23	2	13	1745
Males	11.68	0.47	7	1.29	1	7	1730
T. gracilis, all	11.70	0.65	24	0.71	1	24	1745
Females	11.64	0.65	19	0.86*	1**	19	1745
Male	11.89	0.67	5	0.13	0	5	1800
T. longipes, all	11.66	0.73	28	0.61	1	28	1730
Females	11.59	0.83	20	0.70	1	20	1715
Males	11.84	0.35	8	0.39	1	8	1745
T. montana, all	12.07	0.77	19	0.63	1	17	1800
Females	11.87	0.81	8	0.84	0.5	8	1800
Males	12.21	0.75	11	0.45	1	9	1800
T. tana, all	10.92	1.05	74	0.67	1	72	1645
Females	11.15	0.94	53	0.70	1	54	1715
Males	10.34	1.11	21	0.57	1	18	1600

NOTE: Mean total length of active period (nest to nest); time recorded as resting (inactive for two consecutive triagulations or more) per active period (means of all complete tracking periods including those with no rests); median number of rests recorded per active period; and median time at which treeshrews returned to the nest at the end of activity.
[a] Beginning of 15-minute period into which median falls.
*Probability = 0.06; ** Probability = 0.04 (Mann-Whitney U). Males significantly different from females.

hidden arboreal sites). *T. minor* also sometimes seemed to take refuge from heavy rain in a nest site. Ordinarily, *Tupaia* seemed to stop activity to rest wherever the day's wanderings had brought them, often near a fruit tree that they would revisit later.

The two younger female pentails had a strikingly different behavior from that of any *Tupaia:* on nine of eighteen nights monitored they returned to the nest tree in the middle of the night for long rest periods. The adult female (F181) also did this on one of eight nights that she was followed. The young female F96 returned to the nest tree to rest, dividing her activity into two parts, on five nights between 22 November and 1 December, but on the last three nights she was tracked, between 8 and 30 December, she stayed away from the nest all night. This seems to show that forays out and back to the nest in pentails might be a juvenile behavior that largely disappears with independence. The other young female returned to the nest for midnight rests on four nights, interspersed between five nights during which she did not return to the nest at all.

The distribution of longer rests, detectable by the radio-tracking method I used, is the inverse reflection of activity (fig. 7.2). All species concentrated resting toward the middle of the active period, but only pentails showed a sharp, unimodal, peak of mid-activity rests. These long rests divided the night into two well-defined peaks of activity during the first and last thirds. All the *Tupaia* species were 100 percent active only in the first two to four hours of the morning, and they would rest sporadically almost any time after that. All species have a bimodal tendency for inactivity, with rests most often in late morning and midafternoon. The inverse, overall activity is basically bimodal when viewed cumulatively, especially in *T. longipes,* for which resting was most strongly clustered; but within the midday period there were two minor activity peaks.

These averaged trends do not show what an individual treeshrew does in a day, which is better reflected in the statistics (fig. 7.2, table 7.1). While tracking *Tupaia* on foot from daybreak to dark, matching their movements on a parallel trail, it immediately becomes obvious that they seldom rest. The two statistics, median number of rests per day and mean total time resting recorded per day, together show that the four largest species rest, on average, less than an hour each day and most often only once. That is, on many days no rests at all were recorded. The amount of time spent resting was not significantly different between any species pairs, but the difference between the sexes was significant for *T. gracilis,* for number of rests, and nearly so for time spent resting (with more data it might reach significance). Short rests of less than 20 minutes may have

Activity Patterns

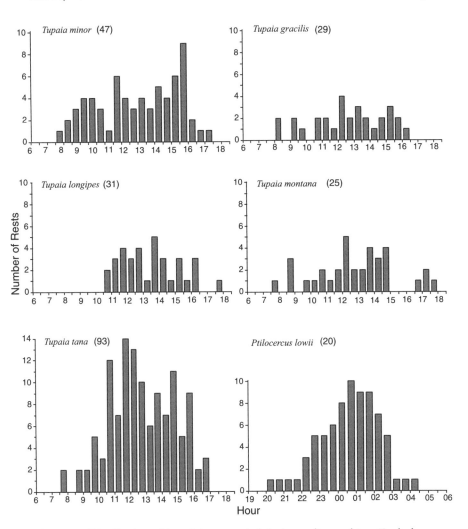

Fig. 7.2. Distribution of inactivity recorded during radio-tracking. Excludes rainstorms. Rest period scored in the 30 min period during which its major part occurred; if an inactive period was longer than 30 min, it is scored into each half hour of which it formed a major part. Y axis = number of times a treeshrew rested in that period. N = number of inactive periods recorded for the species, not the number of 30 min periods in which resting was recorded.

been common, but monitoring of radios with activity sensors showed inactivity only during one or two triangulations per day (of 30–40) that were not part of longer, recorded rests, and the steady movements plotted between points also suggest continuous activity for most days and individuals (see chap. 8).

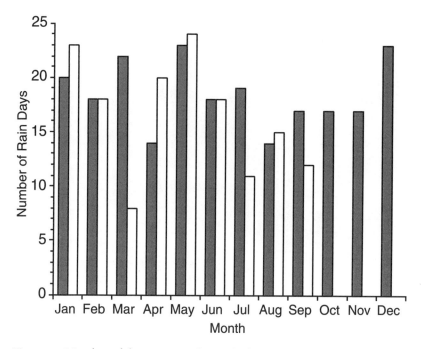

Fig. 7.3. Number of days per month on which it rained at Danum Valley in 1990 (shaded bars) and 1991 (open bars).

RAIN

In the equable temperatures of lowland equatorial forests, rain is the main climatic event that influences animal activity. The pentails foraged near their nest tree during continuously rainy nights and returned to it if nearby at the start of heavy downpours. *Tupaia* species ceased activity during heavy rains, usually after a short movement to a nearby spot, evidently a known shelter. If a storm struck in the afternoon, treeshrews often went to a nest and stayed there for the remainder of the day. Rarely, a *tupai* made straight for a nest and retired when black sky, thunder, and winds presaged an afternoon storm. If it was raining when treeshrews were due to emerge at dawn (or dusk for pentails), they sometimes delayed the start of activity, but eventually they started to move around in moments of lighter rain.

Because rain caused treeshrews to cease activity, I took advantage of the detailed rainfall records from the Research Centre hydrology project

Activity Patterns

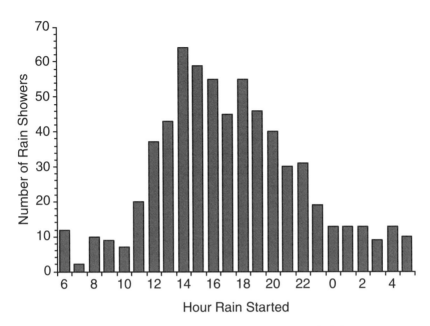

Fig. 7.4. Hour of onset of rain at Danum Valley, all recorded showers.

to estimate approximately how much activity time treeshrews might lose to rain throughout the year.

At Danum Valley Field Centre in 1990, it rained on 222 days, or almost two-thirds of days, and in 1991 it rained on 149 days from January to September (fig. 7.3; see fig. 3.5 for monthly rainfall). In 1990 rain fell for more than 339.77 hours (this is the sum of available records, but data on length of a few rains were lost because of chart-recorder problems or failure to change the paper, although data for the amount of rainfall are complete). Per month, it rained for from 10.92 hours (April) to 56.58 hours (December), or 7.9 percent of the time.

The degree to which rain might truncate an animal's activity depends on which part of the day it occurs. The hours during which most rains began (fig. 7.4) and the hours in which it fell (fig. 7.5) varied somewhat with the season. Throughout the year it mainly rained in the afternoon or first hours of the night. Only in December did rains also start in the last half of the night and continue into the morning. From July through October rains were bimodal, usually starting either near midday or early afternoon or just after nightfall (fig 7.4). In June alone, rain onset was concentrated only in the evenings between 16 h and 20 h. For *Tupaia* species, December was the month during which maximal rain (10.67 h)

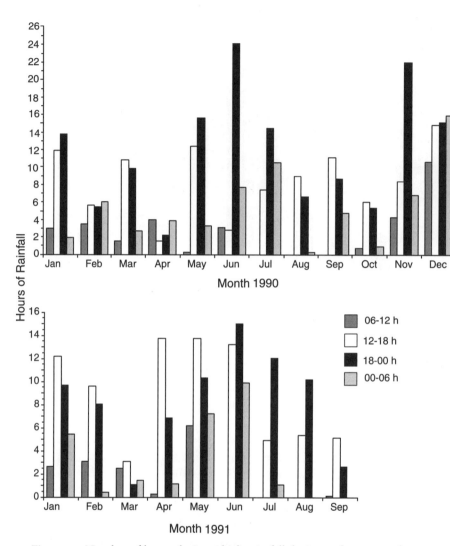

Fig. 7.5. Number of hours during which rain fell during each quarter of the day, for each month of the study. Some records are missing from the data because of chart recorder problems.

fell during their most active, morning period (06–12 h) and during which it rained longer in the afternoon (14.83 h), for a total of 25.5 hours of rain during their potential foraging time. This amounts to only 7 percent of daylight hours for that month. For most months of the year, *Tupaia* species had four to five hours with little likelihood of rain in which to forage following the night's fast. Because it almost never rained in the morning, *Tupaia* species nearly always had several hours in which to feed before rain impeded activity. Lesser treeshrews would become inactive during afternoon rains, and F112 once retired for the day at the mere threat of a storm at 1615 h; but during moderate morning rains in December, M91 and F70 began activity as usual at dawn and moved around slowly in the rain, probably forced to do so by hunger.

Pentails are faced with a high probability of rain at the hour when they become active (figs. 7.4, 7.5). In June and November 1990, 22 to 24 hours of rain occurred during the first half of the night (18–00 h), or up to 13 percent of those hours, and in January, May, July, and December more than 13 hours of rain fell between 18 h and 00 h. At the maximum in June, it rained for a least 32.43 hours during the active periods of pentails. When I monitored pentails during torrential evening rains (from a tarp shelter next to the nest tree), I found that they at first remained inside the tree but were active within it and then emerged during the rain for short forays into neighboring large trees, where perhaps they were able to forage within sheltered nooks or beneath giant branches.

DISCUSSION OF ACTIVITY PATTERNS

The outstanding feature of treeshrew activity is the sheer amount of it. With known rests subtracted, the least active species (*P. lowii*) spent a mean of 8.7 hours each day in motion, while the most active species (*T. montana*) averaged 11.44 hours in motion. However, short rests that were not detected and rests enforced by rain decreased these totals by unquantified amounts. Treeshrews spent little time with others, and as corroborated by direct sightings, most of their activity seemed to be concerned with foraging or feeding, especially in the case of females (see chaps. 8, 9). For most *Tupaia* species, the modal time of entry to the nest was within an hour of nightfall, but the large variation of return times (fig. 7.1) implies that when treeshrews had eaten enough, they retired for the rest of the day. This could be field tested experimentally, by artificial provisioning.

The largest species, *T. tana*, has a shorter period of activity than the others, and the second largest, *T. longipes*, tends to a lesser extent to retire earlier than yet smaller species (table 7.1). This likely reflects the smaller relative metabolic needs associated with larger body size; but in that case food acquisition time would likewise have to be scaled, as absolute needs would be greater in larger species. The much longer activity of *T. tana* females than males is I believe associated with reproductive effort (see chap. 10), likewise fitting a hypothesis that total length of activity is driven by feeding requirements. When a treeshrew has eaten enough for the day, return to the nest would confer safety from avian and felid predators and also decrease daily energy needs with reduced metabolic expenditure during sleep.

Pentails have a relatively shorter total period of activity than do the diurnal species of the same small body size. Since most pentails finally retired only at dawn, their short activity was due both to long midnight rests and to the delay in exit from the nest tree until pitch darkness, half an hour or more after nightfall. The ability to maintain themselves with shorter activity than small *Tupaia* may be the result of a low metabolic rate and torpor while inactive (Whittow and Gould 1976). In captivity West Malaysian pentails kept by Gould (1978) likewise became active quite late (1945–2015 h), suggesting that this pattern is intrinsic to pentail behavior. By rising late, pentails could avoid exposure to crepuscular predators such as owls, which hunt most effectively in dim light.

In a study of *T. minor* in West Malaysia (observations from a fixed site), D'Souza (1972) found that lesser treeshrews began activity very late (0745 h), two hours after the 0545 h sunrise, and then exhibited a sharply bimodal activity pattern with no activity from 12 to 14 h, a peak in late afternoon, and return to the nest at about 1800 h. This contrasts with my data from Sabah, which show exit from the nest universally at dawn and resting taking place any time from late morning to late afternoon, with a slight trend of trimodality in activity. D'Souza saw evidence that after days of heavy rain, lesser treeshrews began activity earlier, presumably to compensate for time lost, and he speculated that late exit from the nest represented slack time that could be used if needed. If lesser treeshrews have shorter active periods and longer rests in West Malaysia than on Borneo, the former site may have higher food availability and reduced obligate foraging time. Unfortunately, the differences in our methods preclude direct comparison. In Sabah it was clear that if treeshrews had "extra" time, they cut activity short at the end of the day, never at the beginning.

The daily need for food must be greatest when treeshrews emerge from their nests following their twelve-hour sleeping phase. For pentails, emergence from the nest in almost all months coincides with the time of day with most likelihood of rainstorms (fig. 7.5). Frequent rains at nightfall could restrict foraging at the most critical time of the night and hence could reduce the ability to acquire extra nutrients for reproduction. Their low metabolic rate may help pentails cope with foraging delays or deprivation caused by rain. Because of their timing, rains at Danum seemed

Activity Patterns

unlikely to have much influence on the time available for feeding of diurnal treeshrews, although in the rainiest months some foraging deficit could occur in particularly rainy weeks.

The direct effects of rain on animals have rarely been studied, and detailed rainfall records are available for few sites. For tropical rainforest birds, it has been suggested that the breeding seasons exclude the rainiest months because heavy rains both physically destroy nests and curtail vital foraging time (Brosset 1990, and pers. com.; Foster 1974). If birds breed during the wet season, rain may reduce the clutch size that parents can feed (Foster 1974).

In summary, the daily activity cycle of *Tupaia* species in Sabah, as monitored by radio-tracking, shows sustained activity and relatively little rest during the day. This implies that foraging takes most of the day. *Ptilocercus lowii* had both a shorter active phase and more resting time than the *Tupaia* but nonetheless foraged nightly for about nine hours. These behaviors imply that the energetic balance of treeshrews may be tight, a hypothesis for which I give more evidence in the next two chapters.

EVOLUTIONARY NOTES

Long daily foraging times in treeshrews are linked to their short, simple digestive tracts and rapid passage times (chap. 5). Food must be ingested repeatedly, in short bouts throughout the daily active phase. In captivity at the National Zoo, *T. tana* from Sabah became obese when fed ad libitum and developed dewlaps and aprons of fat (Miles Roberts pers. com.; pers. obs.), whereas in the wild no body fat was ever noted. In the same captive circumstances, *T. minor* did not become obese. Captive common shrews (*Sorex araneus*) also put on much body fat, but wild shrews are never fat (Churchfield 1990). That wild treeshrews were always lean and foraged (moved actively about) for nine to eleven hours of each twelve-hour day suggests that they need long daily foraging times just to break even energetically. *Tupaia* species sometimes retired to their nests before the end of the daylight period, so rather than more foraging, they opted for shelter, away from the risk of predation by raptors and perhaps other predators. As predicted from theory concerning body size and metabolic rate, the larger the treeshrew species, the more likely it was to enter the nest early.

The smallest nonflying insectivores, the true shrews, are destined by their tiny body sizes and high relative metabolic rates to frequent search for food. Churchfield reported,

Most shrews must feed regularly every two or three hours or they will die of starvation, so they must remain active both day and night. Periods of activity are spent bustling through the undergrowth or along subterranean tunnels as they search for prey and explore their home ranges. Bouts of activity may last from about 30 minutes to two hours, and alternate with periods of rest in the nest. (1990: 123)

Common shrews have about ten bouts of activity per twenty-four-hour day (Churchfield 1990). Treeshrews seem to have a similar, high activity pattern (although with few, or short, rests), but it is restricted to half of the day. Most small mammals, including treeshrews, most tenrecs (Eisenberg and Gould 1970), elephant shrews (Rathbun 1979), marsupials (Flannery 1990; pers. obs.), prosimians (Doyle and Martin 1979), bats, and most but not all rodents (volelike grassland herbivores are among the exceptions) have evolved nocturnal or diurnal lifestyles with twelve-hour activity cycles. This permits these species to tune (by evolution) their physiology and sensory systems to optimal function for their activity phase and to become more efficient at feeding on resources with different day or night characteristics. Cathemeral (active day and night) activity in mammals appears to be found only in species that are unable to acquire enough nutrition within one of the twelve-hour phases, from constraints of either digestive/acquisition systems (grazers), or search time for rare or hard to catch prey (small felids), or metabolic constraints (shrews). Other benefits of twelve-hour activity include shelter from predation for at least half of the twenty-four-hour day and whatever physiological advantages long sleep confers.

APPENDIX: METHODS

We triangulated the location of a treeshrew every ten to twenty minutes, or more often if it was moving rapidly. The number of locations logged per day thus depended both on how many hours the animal was active and on how quickly and far it moved; but most daily samples were of about 40 points. For most of the above analyses, data are included only from the 185 entire days that were not grossly curtailed by heavy rains.

The beginning of activity, when an animal left its nest, was usually precisely signaled by a clear change in intensity or pattern of the radio signal. I often listened continuously to capture this moment. The instant of return to the nest was sometimes also readily identified (as when an animal went underground or climbed to a high nest), but at other times it had to be inferred from cessation of activity for the rest of the day. If a treeshrew went to its nest early in the day, I waited in the field and monitored it until the end of the day, to make sure that it did not resume activity. The beginning of activity was determined on a greater

Activity Patterns

number of days than were completed with a full day of tracking. All records were used in the analysis.

During tracking, we often recorded our perception of whether an animal was active or inactive, and for transmitters with activity monitoring options, the transmitter activity signal was noted. However, for the analyses in this chapter, I do not use these measures but instead define inactivity, or "rest," as any two sequential triangulations between which the animal did not move. This standardizes the treatment of the data and allows more of it to be used, but obviously rests of shorter than twenty minutes were not detected, so that the time spent resting is underestimated. To economize batteries, we turned off the radio receivers between triangulations. For treeshrews with activity sensors, it would have been better to sample some days by running the receiver continuously to measure all inactive periods exactly.

CHAPTER 8

Use of Space

HOME RANGE

A home range estimate is built from a map of the points where an animal is known to have been. The points we know are but a tiny sample of the terrain that the animal might use, and a good deal of thought and theoretical discussion has been devoted to divining the best way to estimate the true home range from isolated points (e.g., the six models summarized in Kenward 1987). Researchers who take radiolocations only every few hours or days do not know how the animal traveled between points, and models that fill in empty spaces, such as "the minimum convex polygon," are often used to connect points for home range estimation. Although following an animal all day, as we did in this study, generates a series of points, we know quite accurately where the animal went, so the points can be connected into a line describing its path. After following many species of mammals by continuous radio-tracking over the years, including squirrels, porcupines, ocelots, and spiny rats, I have come to the conclusion that home ranges, especially when they are also territories, are often highly irregular in outline, with odd hollows and protuberances. Thus, to estimate treeshrew home ranges, I have connected the outer points to form a minimum polygon, without enclosing large peripheral areas where I never recorded the animal to be. The home range outlines are therefore in many cases concave or spiky. The estimated ranges are smaller than they would be with the more standard convex polygon method, but the most important difference between the methods is that the apparent overlap between adjacent territories that is often seen with convex polygons almost entirely disappears with minimum polygons, with consequent implications for social structure (see chap. 9).

Home ranges are spatial projections of all the resource needs of individuals, and they thus hold whatever is needed by an animal at a particular time. For most mammals, it is best to compare only females to address ecological questions about home ranges—how much space is used by members of species with different diets? how far or long must individuals forage? how many can live in a hectare?—or other topics related to energetics. The reasons for this are straightforward: (1) females carry the energetic load of the production of offspring, so only the home ranges of females reflect the ecological baseline for reproduction, or fitness; (2) male travels and home ranges often reflect social, not ecological, motives, because the typical mammalian social organization is such that the agenda of males is not to use spatial resources efficiently but to acquire access to as many females as possible. As a main aim of this study was to try to put the reproductive system of treeshrews into an ecological context, I deliberately biased radio-tracking efforts heavily in favor of females. Moreover, adult males of most species were scarcer than females, and we caught fewer to radio-collar. Strategically, I first put radios on females and then tried to radio-tag the adult males that used the home areas of collared females. Ultimately, we followed twice as many females as males.

The crude home range values for forty-seven treeshrews followed by radio-tracking and one from trapping show high consistency between members of a species and sex (table 8.1). Because the values for animals tracked at Poring are close to those for the same species at Danum, and their behaviors were similar, data from both sites are treated together for species averages. Some treeshrews went off the edge of the trail system, beyond the reach of the receiver, so their ranges are known to be underestimates, while others were followed too briefly, so their ranges are incomplete (table 8.1). The distribution of total home range size as a function of numbers of points shows that for the most part there were adequate samples (home ranges with the most points were not the biggest ones) (fig. 8.1).

The mean home range sizes for species ranged from 1.5 ha to 10.5 ha (table 8.2). When these are related to body mass (fig. 8.2), it is evident that body size is a poor predictor of home range size: among females of the three smallest species, whose body weights differ by only 20 g,

TABLE 8.1. Home range data for 48 treeshrews.

Species	Sex/No.	D max (m)	D min (m)	M/Day	Area (ha)	No. of Points
P. lowii	F96s	268	200	995 (5)	3.17	230
	F163s	210	131	856 (8)	1.93	281
	F181	376	236	1,376 (7)	6.41	288
T. minor	F70	261	103	1,032 (4)	2.01	207
	F112-p	124	109	667 (4)	0.94	155
	F294	193	139	854 (5)	1.61	231
	M91	224	124	866 (4)	1.59	181
	M126-p	251	80	963 (3)	1.27	151
T. gracilis	F67	419	417	1,765 (9)	10.52	466
	F105-p	660	191	1,339 (7)	7.05	330
	F173	500	271	1,452 (3)	9.77	168
	Mfoot	561	344	2,015 (5)	14.71	223
T. longipes	F56	492	275	2,602 (5)	8.78	285
	F66*	457	292	1,708 (3)	8.31	131
	F86s	389	225	1,735 (4)	5.68	225
	F132-p	580	211	1,442 (7)	7.79	216
	F133-p*	590	216	810 (1)	4.07	85
	M64	404	282	2,480 (3)	8.49	120
	M73*	469	409		9.60	54
	M79s	569	138	2,111(2)	7.55	121
	M138-p*	710	253	2,441 (2)	9.15	80
	M287*	638	271	2,711 (1)	9.96	117
T. montana	F148-p*	243	118	1,376 (2)	2.10	123
	F150-p	218	196	781 (4)	2.40	205
	F154-p	396	247	938 (4)	3.09	171
	M143-p	218	177	971 (4)	2.32	185
	M144-p	318	95	916 (4)	1.48	122
	M155-p	235				5
T. tana	F54	272	156	1,054 (3)	2.32	157
	F58	343	191	1,512 (4)	4.02	155
	F65	430	130	1,185 (3)	2.68	127
	F76	421	283	1,024 (7)	5.63	259
	F78*	283	113	854 (3)	1.94	79
	F100	314	256	954 (4)	4.82	127
	F106-p	271	109	897 (5)	2.42	129
	F109-p	335	193	897 (15)	3.97	534
	F166	334	220	914 (3)	3.65	125
	F176	224	156	980 (4)	2.14	141
	F297	227	210	1,006 (3)	2.50	112
	MScar*	320	235	1,584 (2)	4.05	149
	M7	336	221	970 (3)	3.52	128
	M55	424	200	1,327 (2)	6.48	139
	M62	330	247	792 (2)	4.51	75

Use of Space

TABLE 8.1 (continued)

Species	Sex/No.	D max (m)	D min (m)	M/Day	Area (ha)	No. of Points
T. tana	M63	444	338	1,470 (3)	8.04	127
(continued)	M77 *	621	201	1,451 (4)	7.56	158
	M111-p*	455	241	1,363 (1)	6.34	88
	M167	317	199	1,045 (3)	3.31	117
	M168*	276	220	1,218 (1)	2.64	94

NOTE: D max = major axis (maximum diameter) of home range; D min = minor axis. Measured from maps of all compiled location points. -p = animals followed at Poring, (s) after animal number = subadult. M/day is the average of all complete days of tracking (N), excluding days where the animal was lost out of range or tracking was grossly curtailed by heavy rain. (*) animals whose home ranges were partly out of range (incomplete).

TABLE 8.2. Home range data, means for species and sexes.

Species	N	D max (m)	SD	D min (m)	SD	Area (ha)	SD
P. lowii, all (F)	3	284.7	84.3	189.0	53.4	3.84	2.3
Adult female	1	376.0		236		6.41	
T. minor, all	5	210.6	55.1	111.0	22.3	1.48	0.4
Female	3	192.7	68.5	117.0	19.3	1.52	0.5
Male	2	237.5	19.1	102.0	31.1	1.43	0.2
T. gracilis, all	4	535.0	101.6	305.8	97.0	10.51	3.2
Female	3	526.3	122.6	293.0	114.6	9.11	1.8
Male	1	561.0		344.0		14.71	
T. longipes, all	10	529.8	104.3	257.2	70.3	7.94	1.8
Female	5	501.6	84.7	243.8	37.1	6.93	2.0
Male	5	558.0	123.8	270.6	96.4	8.95	1.0
T. montana, all	5	271.3	71.5	166.6	61.1	2.28	0.6
Female	3	285.7	96.3	187.0	65.0	2.53	0.5
Male	2	257.0	53.5	136.0	58.0	1.90	0.6
T. tana, all	20	348.9	94.0	206.0	56.0	4.11	0.4
Female	11	314.0	68.4	183.4	56.8	3.28	1.2
Male	9	391.4	106.9	233.6	43.1	5.16	2.0

one has the smallest home range (*T. minor*, 1.5 ha), another has a home range that is three times larger (*Ptilocercus lowii*, adult, 6.4 ha), and the third is six times as large, the largest of all treeshrew ranges (*T. gracilis*, 9.1 ha). Likewise, the largest species, *T. tana*, has a home range that is smaller than those of three smaller species. Overall, *T. minor*, *T. montana*,

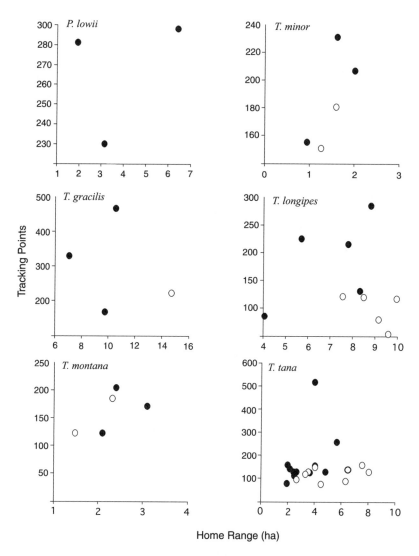

Fig. 8.1. Total number of radio-tracking location points recorded for each treeshrew followed and its home range in hectares. Females = black circles; males = open circles.

and *T. tana* used relatively small areas, while *T. longipes*, *T. gracilis*, and *P. lowii* used relatively large areas. The differences between areas used by the first three are significantly different from those used by the second three (P < 0.05). These home ranges all overlapped on the same study sites and were determined over the same months, so differences

Use of Space

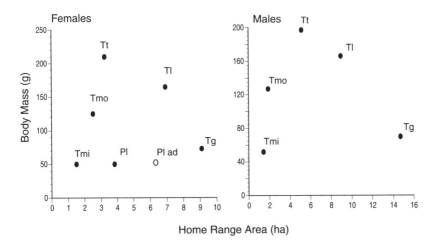

Fig. 8.2. Treeshrew home range areas as a function of body mass, means for each sex. For *Ptilocercus lowii*, the species mean includes two probable subadults; the open circle represents the adult, reproductive female F181.

between them are due only to differences in the way that species behaved and not to habitat or seasonal differences in resources. Males had larger home ranges than females in *T. gracilis*, *T. longipes*, and *T. tana*. These are not significant, primarily because sample sizes for males are too small. Certain males sometimes vanished beyond radio range, so their home ranges were larger than estimated, while this was rarely the case for females.

SPEED AND DAILY DISTANCE TRAVELED

Paths traced by radio-tracking are straight lines between triangulated points, but animals actually zigzag, so recorded path lengths are minima and true distances moved must always be greater. One researcher has measured the difference between the real path of a small mammal and that given by radio-tracking. Guillotin (1982) followed spiny rats (*Proechimys* spp.) by sight and measured their exact itineraries to compare with radiolocation data. He found that the mean real path was 1.8 times longer than the telemetry-estimated one. Thus even "continuous" radio-tracking may largely underestimate path length. In addition, radio-tracking does not register the vertical displacements of arboreal or scansorial species (perhaps equal to or greater than their horizontal travels). Animals thus travel much farther than we can measure by usual field

TABLE 8.3. Mean distances traveled by treeshrews per day.

Species	Distance (m)	Range (m)	SD	Rate,[a] (m/h)
P. lowii, all (20)[b]	1,073	568–1,613	295	124
Adult female (7)	1,376	1,012–1,613	37	131
T. minor, all (20)	871	558–1,234	175	83
Female (13)	851	558–1,234	200	80
Male (7)	908	779–1,057	120	87
T. gracilis, all (24)	1,654	722–2,522	474	151
Female (19)	1,559	722–2,240	457	145
Male (5)	2,015	1592–2,522	382	171
T. longipes, all (28)	1973	810–3,932	702	178
Female (20)	1800	810–3,932	744	165
Male (8)	2407	1,691–2,711	114	210
T. montana, all (17)	958	685–1,510	205	84
Female (8)	859	705–1,056	123	78
Male (9)	1047	685–1,510	230	89
T. tana, all (74)	1078	521–1,930	300	105
Female (53)	1009	577–1,682	256	97
Male (21)	1250	512–1,930	340	128

[a] Rate of movement is crudely estimated from mean m traveled ÷ (mean h active − mean h resting).
[b] ( ) = number of full-day samples.

methods. Even with these likely underestimates, treeshrews traveled impressive distances in their daily rounds (table 8.3). Slender and plain treeshrews were the champion runners, with some individuals regularly logging more than 2,000 m per day.

The rates at which treeshrew species normally moved varied by a factor of about 2, with *T. longipes* and *T. gracilis* in a class by themselves, moving faster than the others (table 8.3). The highest day's average was recorded for *T. longipes* F56, who on 26 October traveled 3.9 km, moving for 11.8 hours at the prodigious mean rate of 333 meters per hour. The rates in table 8.3 are averages throughout the day and all activities. When running quickly, *T. longipes* streaked through the forest at over 800 meters per hour (with a sweating researcher panting in distant pursuit).

Males of all species moved both faster and farther in a day than females, even when their home ranges were slightly smaller (*T. minor, T. montana*).

HOME RANGE SIZE AND ENERGETICS

Home range size has long been used as an index of bioenergetics. Simply put: "The size of the home range in mammals, accordingly, is determined by the rate of metabolism. A large mammal has a larger home range than a small mammal, because it uses more energy and, therefore, needs a greater area in which to find this energy" (McNab 1963: 136). As McNab showed in this early paper, and developed later (McNab 1983), there is a predictable relationship in mammals between body size, metabolic rate, diet class, and home range size. This idea makes intuitive sense and works generally, but within the log-log plots that demonstrate such trends is much scatter that looms large when plotted on linear scales (see fig. 8.2). This scatter may result from differences in diet or social behavior, especially if males are included. If only females are compared, it may result from the fine-tuned ecological differences in foraging habits that comprise the individual adaptations of species, as I shall argue below.

The connection between home range size and actual energy expenditure is indirect (as opposed to metabolic need). The amount of energy that an animal uses for daily activities, above its basal metabolic rate, should be somehow directly related to the actual distance that it travels, along with associated factors such as gait and terrain. The home range area, however, is largely a function of the directions that travel takes, or the shape of the path, and not necessarily its length. To look at this relationship, the mean distance that each *tupai* traveled per day can be plotted against its own home range size (fig. 8.3). This shows that the individuals of a species generally show consistency but that there are marked interspecific differences. The steeper the slope, and the higher the Y intercept, the greater the distance traveled per hectare of home range. Thus, at the extremes, a *T. minor* that travels 1,000 m per day has a home range of 2 ha, but a *T. longipes* that travels that distance has one of 5 ha. If a *T. minor* were to travel 2,000 m a day, its home range would only be 4 ha, but a *T. gracilis* that runs 2,000 m has a home range of nearly 15 ha.

If an animal zigzags, goes back on its path, or returns repeatedly to the same place, it may travel a long way within a small area. In a graphic example from the data (fig. 8.4), on one day *T. minor* M91 traveled 979 m within an area of 0.715 ha; and on another day *T. gracilis* Mfoot ran 1,592 m (his shortest recorded daily path) but covered 7.195 ha. The species means for the distance/area relationship show clearly that as home ranges become larger, daily path length increases at a much slower rate,

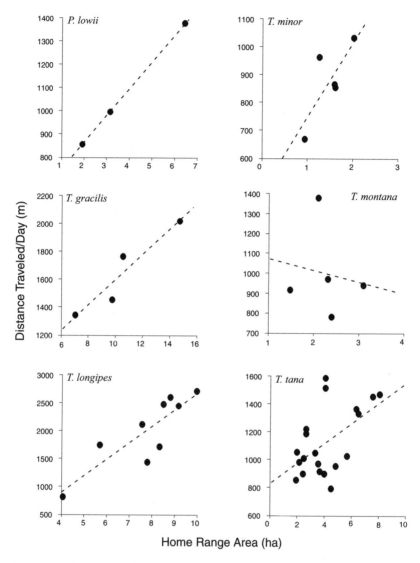

Fig. 8.3. The relationship between mean meters traveled per day and area of home range, for all individuals of each species. Note that scales are different.

such that from one extreme to the other, a doubling of the path length increases the home range area fivefold (fig. 8.5). Treeshrews with smaller home ranges therefore use their land with higher intensity, with more travel per unit area, than do treeshrews with larger ranges, and conversely, an increase in home range size does not entail a comparable increase in

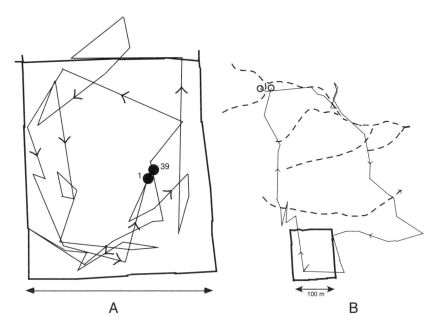

Fig. 8.4. Contrasting patterns of daily use of space, one complete day's movement. A, *T. minor* male M91, 7 December 1990, 39 location points, 974 m traveled, 0.72 ha covered. Square is a 1 ha quadrat of the grid of study-area trails, circles are nest sites. B, *T. gracilis* Mfoot 31 July 1991, 41 location points, 1,592 m traveled (one of his shorter days), 7.20 ha covered. Square at bottom is the same quadrat shown in A. Heavy dashed lines are stream courses.

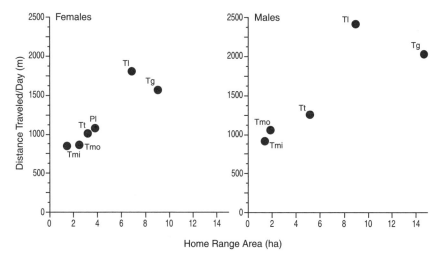

Fig. 8.5. Mean distance traveled per day by treeshrew species as a function of mean home range size, for each sex and species.

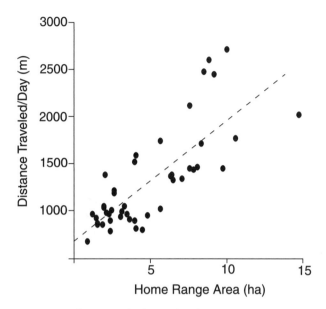

Fig. 8.6. The general relationship between mean meters traveled per day and area of home range, from data for all individual treeshrews (linear regression: m = 667.52 + 129.25 area).

travel expenses but a lesser one. As figure 8.6 shows, the general linear regression for all species and individuals is

meters traveled daily = 667.52 + 129.25 × Area (ha).

Thus, although the home range of a female is a map of the distribution of the resources that she needs/uses during a particular time span, its area reflects the configuration of the daily foraging paths more than their absolute lengths. For males, the relationship between distance traveled and home area was similar to that for females of the same species (see fig. 8.5).

On the maps of daily movements, I also measured the area of the daily range, but it was evident that this calculation was both useless and misleading as an ecological descriptor: the real land use by an animal on a particular day is the narrow strip along its actual path. Daily ranges often were odd shapes, long and thin or curved. Connecting the points into a polygon always included areas that the animal went nowhere near, and deciding how to connect the points was highly subjective. The shape of the path, not its length, determined the daily area.

Another set of measurements often used in animal studies is the maximum and minimum diameters of the home range (see tables 8.1, 8.2). I

Use of Space   135

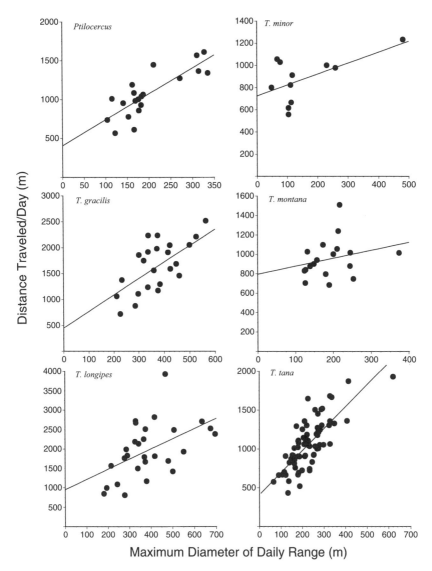

Fig. 8.7. The distance traveled during a day's activity as a function of the maximum diameter of the range for that day. Each point is one individual daily record.

measured these parameters for each daily itinerary. Unlike daily range area, the length of the major axis is directly derived from the linear path of the animal. For some species, the relationship between the distance traveled (i.e., actual ground covered) and maximum diameter is quite good (*T. tana*, fig. 8.7), but for others it is poor (*T. minor*, *T. montana*).

PATTERNS OF TRAVEL AND DESTINATIONS

All *Tupaia* species ate fruit and invertebrates, and I assume that the daily travels of females were mostly directed toward finding one or the other of these commodities. Because I could rarely watch treeshrews, I had to assume that what I did see was typical of the overall activity that was hidden and infer from treeshrew movements what they might have been doing.

FRUIT TREES

Fruit trees were the most evident specific destinations of treeshrew travels. A favored tree was visited repeatedly and could be the focal point of activity (chap. 4). *Tupaia minor, T. gracilis,* and *T. longipes* would intensively concentrate their activity around such trees. Because of their typically vast daily excursions, a focus of movements to a single destination at a fruit tree was most easily detected and unambiguous for *T. gracilis* and *T. longipes* (see fig. 4.1). *T. tana* were rarely so focused, but nevertheless they clearly aimed their itineraries to some fruit sources. *T. montana* was followed too briefly to evaluate its fruit foraging.

For *T. minor, T. gracilis,* and *T. longipes,* the use of fruit trees was in a loose way inversely correlated with the distance traveled in a day (table 8.4). There is too much variation for calculation of mean values, and there are few strictly comparable data sets (the same animal during the same month with and without a fruit tree), so I simply give the raw data. This shows that treeshrews usually traveled less far on days when they spent the most time at known fruit sources. This result can be viewed in two ways: either the time spent at fruit trees simply made less time available for other travel; or when fruit was an important food source, the distance required for foraging was smaller. One or both may be the case on particular days, but the result would seem to be that because treeshrews traveled less distance per day when feeding on some fruit species, they used less travel energy per day. The distance "saved" by eating fruit was up to 30 percent, compared to movements when no focal fruit tree was apparent.

OTHER FEATURES OF THE ENVIRONMENT

Treeshrews used the entire forest but preferred some areas over others. All terrestrial treeshrews at Danum Valley that had it in their home range spent

TABLE 8.4. The relationship between time spent at fruit trees during a day, the number of visits to the tree that day, and the total distance traveled per day by treeshrews. The data are grouped by two-month seasonal periods. Only use of known fruit trees is included; each line is the record for one day. nr = none recorded.

Treeshrew	Month	No. of Visits to Fruit	Hours at Fruit Tree	Distance/ Day (m)	Fruit
T. minor					
F112	April	7	6.44	678	Myrsinaceae, *Ficus*
		5	4.74	667	Myrsinaceae
		7	3.05	824	Myrsinaceae
		3	2.16	617	Myrsinaceae
		3	2.11	558	Myrsinaceae
M126	May	7	4.65	666	*Ficus*
		3	2.31	802	*Ficus*
		nr	nr	1,057	none
		nr	nr	1,030	none
F70	October	4	1.24	913	*Parthenocissus*
		2	0.91	979	*Parthenocissus*
		1	0.62	1,234	*Parthenocissus*
		nr	nr	1,048	none
		nr	nr	1,002	none
T. gracilis					
F105	May	1	4.21	722	*Ficus*
		1	nr	878	*Ficus*
		2	nr	1,060	*Ficus*
		nr	nr	1,297	none
	August	3	1.72	1,671	*Ficus*
		2	1.10	1,174	*Ficus*
		1	1.00	1,461	banana
		3	0.83	2,051	*Ficus*
		2	nr	1,793	*Ficus*
F67	September	2	2.60	2,240	*Dimocarpus*
	October	3	4.84	1,982	*Dimocarpus*
		2	0.93	1,858	*Dimocarpus*
		4	0.48	2,238	*Dimocarpus*
F173	July	4	2.6	1,376	*Alangium*
	June	nr	nr	1,919	none
		nr	nr	1,458	none

TABLE 8.4. (continued)

Treeshrew	Month	No. of Visits to Fruit	Hours at Fruit Tree	Distance/ Day (m)	Fruit
T. longipes					
F132	May	1	5.80	997	*Ficus*
		4	4.17	1,816	*Ficus* (2 trees)
		6	3.84	1,669	*Ficus* (2 trees)
		1	3.60	1,422	*Ficus* (2 trees)
		2	0.73	1,170	*Ficus* (2 trees)
		nr	nr	1,933	none
F133	May	2	2.48	810	*Ficus*
		4	2.14	1,118	*Ficus*
		nr	nr	1,026	none
F56	October	3	0.86	3,932	*Polyalthia*
		2	0.17	2,821	*Polyalthia*
		nr	nr	2,721	*Polyalthia*
F86	March	5	1.51	2,189	*Ficus*
		nr	nr	2,137	none
M64	December	4	0.50	2,512	*Dialium*
		4	0.46	2,673	*Dialium*
		4	nr	2,255	*Dialium*

many hours in a section of riverside floodplain that had low, dense canopy vegetation and black, wet, claylike alluvial soil laced with worm casts. I never found any major fruit sources there, but it may have been a rich source of invertebrate prey. I could not discover the reason for its attractiveness.

Tupaia tana had a characteristic habit of foraging along the banks of streams (fig. 8.8). This pattern was particularly marked on mornings after drenching nocturnal rains, when the animals would follow a stream closely, working their way along extremely slowly and steadily. For example, in the northwest leg of the path shown in figure 8.8A, F100 took an hour to travel 150 m. Only large treeshrews regularly behaved in this way, and I conjecture that they were hunting earthworms or crabs and crayfish (see chap. 5). Other terrestrial species often used watercourses as travel routes that they followed at normal speed. The paths of treeshrews often touched on streams, which they may have visited to drink water. At Poring the study area was largely on a ridge top without surface water, and all of the radio-collared *T. tana* frequently descended to

Use of Space

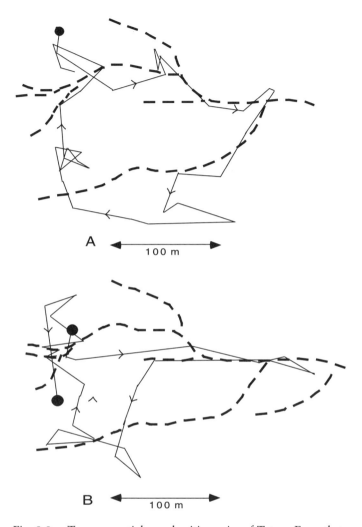

Fig. 8.8. Two sequential one-day itineraries of *T. tana* F100 that show use of stream courses (dashed lines). *A*, 13 June 91, after rain, 33 locations, 1,035 m traveled. On this day she ceased activity at only 1514 h. *B*, 14 June, 28 locations, 1043 m, activity ended at 1424 h. Note her use of different stream sections on the two days.

the bottom of one of the adjacent ravines, where streams always ran. F109 often spent most of the day deep in the ravine, but when she was tending her young, which stayed on top, she would purposefully dash down to the bottom and immediately run back up. She likely descended to drink, because the event was too rapid to be a foraging run. The lactating female pentail went straight to a stream 150 m away when she left her nest on about half of the nights she was followed. The other arboreal treeshrew, *T. minor,* was not noted frequenting watersides.

Treeshrews had particular pathways where they preferred to travel, and different terrestrial species used the same ones. All *Tupaia* species were attracted to treefalls and dense, dark, viny thickets, especially when these were in the damper forest understory, in small ravines and gullies. Treeshrews do not like to cross open spaces on the ground, so if there was a small log, root, or branch across a stretch of otherwise cleared trail, every individual would use it. Likewise, a dark area of low canopy over a trail was used in preference to a bright, high-canopy stretch. *Tupai* also liked to run down the tops of old fallen logs, especially in dense thickets, or down tunnel-like vine tangles along streams. Their pathways thus tended to follow zones of denser cover that might offer refuge from aerial predators and midday heat and/or include enriched invertebrate prey. Figure 8.4B shows the slender treeshrew male following a portion of the identical path down the stream course followed by *T. tana* F100 in figure 8.8A.

Arboreal pathways are more limited than terrestrial ones, and both lesser and pentail treeshrews seemed to have some fixed routes that were probably determined by where adjacent trees connected. Lesser treeshrews seemed to spend the most time in dense lianas that also had many leaves, that is, those with foliage exposed to the sun around trunks of emergent trees, in treefall gaps, on hillsides, or in disturbed forest, but the treeshrews mostly kept hidden beneath the leaves, inside the vine blankets. To cross open trails, they also kept to places with the densest, darkest aerial corridors. No such tendency was evident in pentails, which traveled everywhere and spent much time on bare trunks or exposed in the top of treelets. These differences are likely due to nocturnal, compared to diurnal, activity.

Diurnal treeshrews of different species thus had much in common in that they fed at the same fruit trees and preferred to travel in similar microhabitats. There were nonetheless large differences in the sizes of their home ranges and the distances they moved during a day. These differences must be linked to contrasts in the more species-specific parts of their ecologies: their dietary differences in invertebrate prey and the prey densities and prey encounter rates by foraging treeshrews.

DENSITY AND BIOMASS

Density of individuals is the combined product of home range size and social organization (see chap. 9). I estimated density by measuring the

Use of Space 141

TABLE 8.5. Estimated treeshrew population densities.

Species	Estimated Density (no./km²)			Biomass (kg/km²)
	Average, 1990–91	September–December 1990	March–September 1991	
P. lowii	57			2.96
T. minor	120			7.80
T. gracilis	13	10	16	1.17
T. longipes	23	20	26	5.45
T. tana	49	44	54	10.78
T. montana	84			12.01

NOTE: For *Ptilocercus*, density is calculated from the known area used by the colony at Danum times an estimated four individuals in it. For *Tupaia* species, densities are estimated for breeding adults only. If all females have young, the estimated number doubles. Values for all species except *T. montana* are based only on data from Danum Valley. Estimates are calculated directly from the areas used by radio-tagged individuals of each sex. For species whose separate densities were calculated for different parts of the year, the left column is the average of the two. Biomass is estimated from the densities in the left column and the weights of the actual treeshrews from which the density is calculated. If young are present, biomass might increase up to three-fourths more.

area occupied by an assemblage of neighboring radio-collared individuals (table 8.5). For pentails, there are data only on a single group, and I do not know either their exact number or whether they overlapped with neighboring groups, so the estimate shown is the minimum number that used a known area. For pentails only, the density estimate includes young. The only breeding female used 6.4 ha, so possibly only about fifteen adult females (31 adult animals?) lived in 1 km². The number of young *Tupaia* on the study area varied dramatically throughout the year but always returned to the minimum of none. To estimate the ecological baseline densities for *Tupaia* species, I therefore give the number of resident, potentially breeding adults, without young. The maximum seasonal density for the year would be two to two and a half times the given number, and the average would be about 20 to 30 percent more (table 8.5).

Montane and lesser treeshrews had the highest densities of individuals. In the lowland syntopic treeshrew assemblage, there is a striking phenomenon: among the three terrestrial species, density is directly related to body weight, rather than inversely as expected from classical ecological theory. In fact, density approximately doubles with each species step, so that there are about four times more large treeshrews than slender treeshrews. The difference between numbers of the tiny arboreal lesser treeshrew and the tiny terrestrial slender treeshrew is almost an order of magnitude, hinting at pronounced ecological contrasts.

DISCUSSION OF HOME RANGE

The only other published home range data on treeshrews were collected by observation of marked *Tupaia glis* on Singapore (Kawamichi and Kawamichi 1979). This showed mean home range areas of 1.02 ha for males (N = 16) and 0.88 ha for females (N = 18). These are about eight times smaller than those of the comparable plain treeshrews in Sabah. Their "social" density, calculated in a way similar to mine (e.g., adults only), was 240 individuals/km^2 without young and up to 720 with young (Kawamichi and Kawamichi 1982). This correlates with the small home ranges. Langham (1982), in a trapping study of *T. glis* in peninsular Malaysia, also reported tremendously high densities of 369 and 478 treeshrews/km^2. Dans (1993) reported densities of Palawan treeshrews of 1.6 to 3.2 individuals/ha, in the same range as those of Bornean species. No other measurements of treeshrew ranging behavior have been reported.

West Malaysian *T. glis* thus reach numbers tenfold higher than those of *T. longipes* on my study areas in Sabah. In West Malaysia *T. glis* is the only terrestrial treeshrew, so one can ask whether *T. longipes* numbers in Sabah are depressed by competition with other species. Even the sum of the highest recorded densities of all three syntopic terrestrial species at Danum Valley, 96/km^2 (table 8.5) is much less than half of those reported for *T. glis* alone on the mainland. A likely alternative hypothesis is that food resources are more scarce and restrict population sizes on Borneo (see chap. 12).

Although there is little other information on daily ranging behavior with which to compare Bornean treeshrews to other congeners, it is instructive to compare the ranging behavior of treeshrews to that of other mammals. Primates are by far the best studied taxon in this regard, and the data for many genera and species have been summarized (Smuts et al. 1987). Because they are of a size similar to treeshrews and share some ecological features, squirrels (Emmons 1975, 1980; Payne 1979) and elephant shrews (Rathbun 1979) can be usefully compared (table 8.6). Data were collected in different ways by different investigators, and intraspecific variation is common, so species values should be viewed only as ballpark trends. Two features stand out in this small list: (1) for small primates the day range is small relative to home range size; and (2) with

TABLE 8.6. Ranging behavior of other small, insectivorous/frugivorous equatorial rainforest taxa. Mean home ranges of males and females arboreal (A), terrestrial (T).

Species	Mass (g)	Home Range (ha)	Day Range (m)	Movement Rate (m/h)	Reference
Elephant shrews (all T)					
Rhynchocyon chrysopygus	540	1.7			Rathbun 1979
Elephantulus rufescens	58	0.34			Rathbun 1979
Squirrels					
Ratufa bicolor (A)	1,442	3–7	315	30	Payne 1979
Callosciurus notatus (A)	227	<1			Payne 1979
Protoxerus stangeri (A)					
(2 subadult females)	488	4.1	572		Emmons 1975
Epixerus ebii (T)	577	17.5	857	130	Emmons 1975
Heliosciurus rufobrachium (A)	375	4.6	519	61	Emmons 1975
Funisciurus pyrrhopus (T)	330	3.4	343	48	Emmons 1975
Funisciurus lemniscatus (T)	139	1.2	393	47	Emmons 1975
Primates (all A)					
Cebuella pygmaea	115	0.4	290		Soini 1988
Aotus trivirgatus	800	9.2	708		Wright 1985
Callicebus moloch	800	6.9	671		Wright 1985
Saimiri sciureus	900	>250	1,500?		Terborgh 1983
Saguinus fuscicollis	400	30–100	1,140–1,590		Goldizen 1987
Galago alleni (females)	265	10			Charles-Dominique 1977
Galagoides demidoff (females)	61	0.6–1.4			Charles-Dominique 1971
Tarsius spectrum	120	1			MacKinnon and MacKinnon 1980

the exception of one terrestrial squirrel, the widest-ranging species are leapers and vertical-cling-and-leapers that eat many insects (but not all such species), including galagos, tamarins, and squirrel monkeys. In comparison, the salient character of the treeshrew daily ranging behavior in Sabah is that it is greater to much greater than that of most other species in the same weight range. Home ranges are also on the large end of the spectrum. In both home range and distances traveled, treeshrews resemble the most mobile of primates. Because feeding at fruit trees reduces daily path lengths in both treeshrews and tamarins (Terborgh 1983), long daily excursions are most likely due to either search for insects and/or search for undiscovered fruits or widely dispersed small fruit sources. The champion travelers, *T. gracilis* and *T. longipes,* exceed the largest daily distances of much bigger primates. In contrast, the arboreal lesser treeshrew has a home range exactly the same size as that of the lesser bushbaby (*Galagoides demidoff*) of the same weight.

The long daily path lengths of treeshrews go hand in hand with the extended active periods described in the previous chapter. To meet their daily needs, a long time spent moving translates into a long distance moved. Together these patterns are evidence that the food supply of Bornean treeshrews is highly dispersed, so that each animal needs a long daily time and distance to fulfill its basic needs.

CHAPTER 9
Social Organization

Social organization can be broadly defined as the pattern of interactions between individuals. This pattern determines ecological questions, such as how the physical space of the habitat is used by different classes of individuals and how resources are allocated among them; and also more behavioral and evolutionary questions, such as which individuals produce offspring and what happens to those offspring as they mature. Because treeshrews could not often be watched, I had to infer their social relationships from the deployment of their home ranges and the relative movements of radio-tagged and trapped individuals. The key question of whether treeshrew home ranges also constituted territories had to be judged indirectly. As I could not see territorial defense behaviors, I define territoriality simply as the exclusive use of a space by a class of individuals, to the exclusion of others of the same class. If a space were not defended in some fashion, I assume that other individuals would use it.

SOCIAL RELATIONS OF TREESHREW SPECIES

PTILOCERCUS LOWII Lim (1967) reported that pentails nest in male-female pairs, and Muul and Lim (1971) found up to five pentails in the same nest, but they did not report their sexes or ages. Four pentail specimens collected by Muul and Lim on the same day in the same locality, preserved in the National Museum of Natural History (USA), include an old adult male and an old adult female of about 50 g each and two subadult females of 40 g each, just replacing their milk premolars. There are no notes stating whether these were collected together in a nest, but it seems likely. Lyon (1913) likewise cites a record of a male and female captured together when a tree was felled. The membership of nesting groups is thus not well defined, but all available data concur that an adult pair and their offspring nest together.

The first two pentails captured at Danum Valley were nonparous females with unworn, but complete, adult dentition (F96, 43 g, Nov. 1990; F163, 53 g, Mar. 1991), and within the weight range cited for adults (40–60 g; Gould 1978). These were followed by captures of a young adult male (50 g) with slightly worn teeth (July 1991) and, finally, of an adult, larger, lactating female with well-worn teeth (F181, 63 g, Aug. 1991). Each was captured after the radio of the previous one had quit (although transmitters exceeded specifications and ran for 90 to 100 days). The radio on the male slipped off in the middle of the first night of tracking, so I logged only a few data points. This species was difficult to trap, and only one animal was recaptured (once).

These four pentails denned by day in the same hollow tree (see chap. 6) and were members of the same group. The old female F181 was likely the mother of the two subadult females, but the male could have been either her son or her mate. The nest tree had multiple exits, and I counted up to four "large" pentails emerging from the tree at nightfall, but more often I saw only one to three. Pentails whisked at high speed around and around and up and down trunks and vines, materializing and vanishing unpredictably, so it was impossible to be sure of a count unless all were audible/visible at once. Radio-tagged pentails often left the tree out of sight, from hidden canopy holes, and other individuals must have done likewise. Thus it was impossible to ascertain how many were living together, but there was always more than one in the tree.

When they left the nest tree at nightfall or returned to it at midnight or dawn, the pentails often briefly ran about together, one following another around the trunk of the nest tree and through the adjacent mid- and understory vegetation. At these times they were likely to chirp with birdlike trills. Some calls were evidently in alarm, directed at me, and these were most often elicited near the nest tree when a number of pentails were about. Several often emerged together, and they returned from foraging quite synchronously. In contrast, of more than forty observations of pentails away from the nest tree, all were of solitary animals, and all individuals radio-tracked always foraged alone. Pentails therefore socialized at home, in or beside the colony tree, but foraged solitarily.

Social Organization 147

The members of the group of pentails had a remarkable arrangement of home ranges. The two young females had home ranges mostly within the larger home range of their presumed mother, as is usual among mammals, but the two youngsters used ranges almost entirely exclusive of each other, like opposite spokes of a wheel from the nest hub. The male, in his half-night record, went straight to an unused sector between the ranges of the young females but also within the area of the old female (fig. 9.1). The parous female in turn spent most of her time in areas far from the nest, outside those used by the young, although one small zone was heavily used by young F163 and the mother (fig. 9.1). Because I cannot be certain that the previously radio-tagged pentail was still one of those in the colony when the next one was followed (although I believe that they usually were), this intriguing pattern needs further confirmation.

While tracking a pentail, I occasionally saw another, but there was no way of knowing if these were from the same or different groups. I saw no chases. The apparent repulsion between ranges of the young hints of exclusive foraging domains for each individual, within a group domain outlined by the home range of the adult female, but I have no data on whether the colony itself, or the adult female, was territorial. When Gould (1978) mixed strange adult pentails together in a room, a male and female did not fight, while two females at first were aggressive and fought with biting but eventually nested together.

TUPAIA MINOR   Although *T. minor* was the most commonly seen species, we captured relatively few: seven at Poring, of which four were adults, and eleven at Danum, of which ten were adults. At Poring we radio-tagged one adult female (F112). A young adult male with unworn teeth that was radio-collared in her home range (M126) promptly moved to a separate area, and he was probably a dispersing young. At Danum I radio-collared a pair (F70, M91) and, eventually, a neighboring female (F294). Several others were fitted with bead collars, and these, or juveniles with tail-hair clips, were spotted from time to time on both study sites.

The pair of *T. minor* at Danum Valley used almost perfectly coincident home ranges (fig. 9.2). That of the male was slightly smaller than that of the female, and entirely within it. The male was captured two months after the female, when her radio was failing, so their interactions were recorded only during his tracking sample. These two animals met often, usually three to four times a day, and on 7 December they were near each other most of the day while they traveled 980 m, with F70 recorded close to M91 thirteen times. They foraged together for an hour

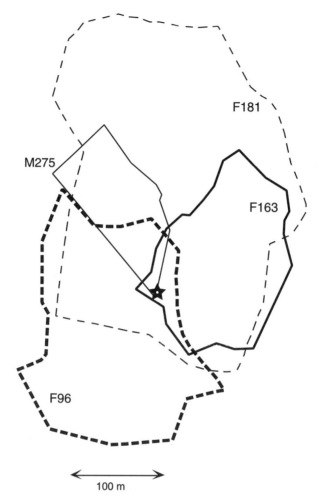

Fig. 9.1. Home ranges of the four pentail treeshrews. Star = the communal nest tree. M275 was followed for only half a night. Note separated ranges of F96 and F163, young females assumed to be littermates, although not followed simultaneously (see text). F181 was an adult, lactating female and the presumed mother of the other females.

in the morning, two hours at midday, and again for another two hours at the end of the afternoon. On another day they met nearly as often. When they foraged together, they were always several to many meters apart, in different parts of a tree or in adjacent trees. Lesser treeshrew pairs foraged in parallel, as would be expected for insectivores, not close in tandem, but a few times one followed another along the same path.

Social Organization 149

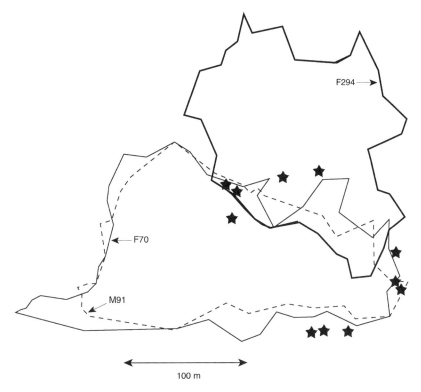

Fig. 9.2. Territories of a pair of *T. minor* (male M91 and female F70) and a female neighbor (F294, heavy line). Stars are locations where unmarked *T. minor* were seen during radio-tracking of members of the pair. At that time F294 had not been captured and she could have been among them.

Treeshrew groupings that I saw throughout the study, including while radio-tracking (table 9.1), show that lesser treeshrews were the most likely to be in pairs. Almost a third of individuals were noted in general proximity to another *T. minor*, but because they were not usually close together, associations were not always clear. The male M91 appeared to be monogamous, because he was restricted to the range of F70, but his female's range may have overlapped those of other males as it was slightly larger.

When a pair of *T. minor* rested in the same fruit tree, they were always far apart, just as they were when foraging. I saw a male sniff and mark and then use the favorite resting perch of F112 in a fruit tree, where she usually rested several hours each day. Afterward she seemed to stay away from this spot, although she did continue to rest there from time to time.

TABLE 9.1. Groupings of treeshrews.

Species	N	Solitary	Twos, Total (%)	Two at Fruit	Two Chase	Female + Young	Other Pairs
T. minor	176	128	24 (27)	7	1		16
T. gracilis	74	64	5 (14)			1	4
T. longipes	76	70	3 (8)	1			2
T. tana	107	95	6 (11)		1		5

NOTE: Number of sightings of single *Tupaia* or two seen fairly close together in time and space. Some of these may not have been associated. N = number of treeshrews; numbers in other columns are the number of observations (for twos, there are twice as many animals as observations). Fruit = seen at a fruit tree; therefore association may be coincidental. Chases were agonistic. More than two were never seen together. Includes treeshrews seen while radio-tracking but not *T. tana* F109 with her young.

The two neighboring females, F70 and F294, had almost exclusive ranges, and I consider them to have been territorial. They overlapped on the northeast section of F70's range, but it is noteworthy that the area of overlap was not used by F70's male, M91 (fig. 9.2) and that F70 only used this area slightly. Merely 14 of 207 recorded radio-tracking points were within it. These were recorded in October 1990, and she was not registered on the area in December when M91 was followed. F294 was not followed until much later, in June 1991 (when F70 was still present but without a radio), so it is not certain that there was any simultaneous home range overlap. The only aggressive interaction that I saw between lesser treeshrews (unidentified individuals) was exactly in the corner of that trail quadrant where the two female ranges appear to overlap, evidence that the sector was in dispute. The sites where I saw unmarked lesser treeshrews while radio-tracking M91 (stars in fig. 9.2) were all outside the periphery of his and his female, F70's, home ranges, which supports the hypothesis of territoriality.

TUPAIA GRACILIS Slender treeshrews were the rarest of the *Tupaia* captured; only three adults and four or five subadults (one escaped) were trapped on the East Ridge at Poring, and five adults and four subadults were trapped at Danum. At Poring, the home range of the single female collared (F105) was 660 m long and 191 m wide and covered the whole study area. No other adult female was captured, so her range seemed to be exclusive. One male was captured and radio-tagged on her range and was recorded meeting with her, but some mishap occurred, because before he could be followed, his signal became fixed at the bottom of a ravine off the study area, implying that he had died. When the area was

Social Organization   151

next trapped, another adult male was captured. Only one adult of each sex apparently simultaneously used the Poring area.

At Danum, only a single adult, female *T. gracilis* F67, was trapped on the study plot in the first six months from September to March, but I sighted at least one other animal, which could have been the one subadult that was also captured. In late March another female (F173) and two males appeared (M175, Mfoot). All were radio-collared, but the signal of M175 was never received, he was not recaptured, and he may have been a transient. Mfoot was badly injured in the trap and snapped his hind leg completely across above the heel. I immediately released him as he was, without marking or handling, and was amazed and delighted when we recaptured him several times, gradually healing, during the next months. His leg had healed by June, but the foot had lost most toes and was deformed and nearly useless. I radio-tagged him in June and followed him through August. This astonishing three-legged male, of only 80 g, had the largest home range recorded in a treeshrew and nearly the largest daily movements (see table 8.1).

Slender treeshrews appeared to be territorial, with only one adult of each sex present on a home range. The male M175, whose radio signal was never found, was captured on the north border of Mfoot's home range. Mfoot was last seen on 23 August, and on 24 September, the last day of trapping, another very young adult male was caught about 100 m inside his home range on the south side. The territory of Mfoot included the territories of both females F67 and F173 (fig. 9.3) and also extended to the west beyond these, and thus it could have included part of the range of another (but the trapline passed through that area, and none was caught there). The female F67 occupied the southern half of the study area from September 1990 to 19 March 1991. F173, on a day in April, used the contiguous northern half of the study area. Gradually, through April to June, F173 moved south. F67 stayed farther and farther south, until in her last records in mid-April, when her second radio quit, she was altogether south of the study area. F173 clearly displaced F67 in a gradual process lasting about a month. In her last tracking record, on 11 July 1991, F173 used much the same area used by F67 on 24 September 1990, so replacement seemed total. I do not know if F67 was then still alive.

The male *T. gracilis* and F173 both had functional radios when he was followed (but F67 did not). On every day he spent time in the territories of both females, looping from one to the other. On one of the five days he met F173, and his signal followed hers for 40 min as she moved; on

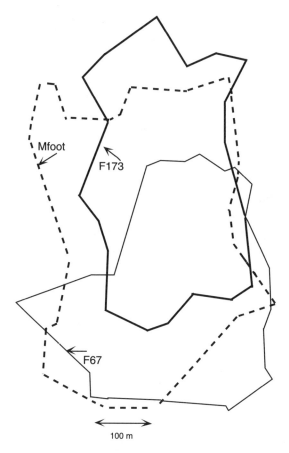

Fig. 9.3. Complete ranges of the three *T. gracilis* followed at Danum Valley. The two females, F67 and F173, did not spatially overlap in time, but F173 took over the zone of apparent overlap, displacing F67 toward the south (bottom), while the male (Mfoot) stayed in place, overlapping both females.

the others they met briefly, for 15 min or less. The male's movements formed giant loops that outlined his borders and rarely backtracked (see fig. 8.4B), whereas the female often went back and forth locally (see fig. 5.1C). The male appeared to be the active agent in encounters, and he went directly to, or past, areas where F173 was to be found.

Slender treeshrews foraged solitarily and were rarely seen together (table 9.1). Radio-tracking showed that the male met the female F173 on most days, and he also went to where he would have found F67, so

Social Organization   153

that social interactions were regular. This male was actively polygynous, with two females. One of these females initially had a home range partly outside of his, in the area in which another male was caught that month, but she eventually moved south entirely into his home range, displacing another female to do so.

TUPAIA LONGIPES   On the East Ridge at Poring, only four plain treeshrews were captured, one subadult and three adults, two females and a male. All adults were radio-collared. At Danum Valley I captured twenty-six individuals: four adult males, five adult females, and the rest young, some of which grew to adulthood on the study area.

Tupaia longipes appeared to be strictly territorial; there was exact meeting but no overlap between the ranges of neighboring adults of the same sex (figs. 9.4, 9.5). The males had territories with most borders almost superimposed on those of the females, but in two of the three pairs for which males and females were simultaneously radio-tracked, the males used some extra territory. At Poring M138 went on a single excursion into an area well outside of the range of his mate, F132. Possibly this "annex" was not defended territory. He did not overlap at all with the neighboring female on the other side (F133). Territories were so large that only one fit on the main study area at Poring, and only two fit at Danum. I never was able to capture the adults that must have occupied the northwest side at Danum, although I trapped their presumed young. The northeast animals used such a vast area that they went out of receiver range and I would "lose" them. Because it was so difficult to get good data with complete daily records, I followed them little.

For one month, I radio-collared a subadult/just adult male, M79, which was trapped in the middle of the study area (Oct.–Nov. 1990), to see how he interacted with the resident adults. His peculiar, L-shaped home range fringed that of the adults, with clear avoidance of their territory (fig. 9.4). He entered their territory seemingly to visit the baited trap sites. He was never recaptured after his radio was removed in November.

In January 1991 a juvenile female (F86) born on the study area in September, and first captured in October (115 g), settled on the west side of the territory of F56. Because she was captured as a juvenile 150 m north of F56's territory, she was probably not a daughter, but the possibility cannot be excluded. This female appropriated almost half of F56's old territory, and F56 then used only the east side (fig. 9.5). This pattern persisted until the end of the study. I once saw these two females meet, and one chased the other, but I could not see which. The male of that territory

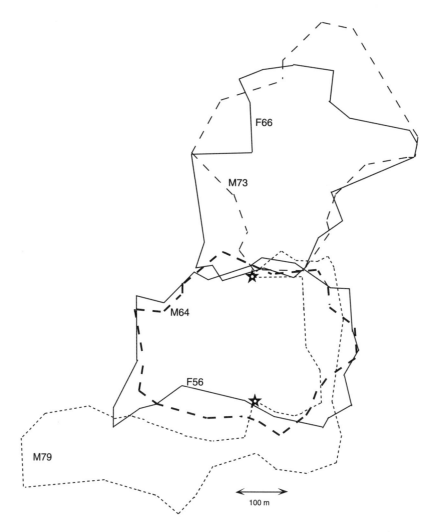

Fig. 9.4. Territories of two neighboring pairs of *T. longipes* at Danum Valley, September to December 1990, and the range of a subadult male, M79, during one month, October–November. The two stars indicate the location of trap sites that apparently drew M79 into the territory of M64.

(M64) became trap-shy, and I never captured him again after March. However, I saw him from time to time on F56's range until the end of the study. I do not know if he still used the area taken over by F86. Both of the only other pair of plain treeshrews on the study area were last captured in December. By March another male (M287) was resident in that territory, but no other female was captured there until the end of August

Social Organization 155

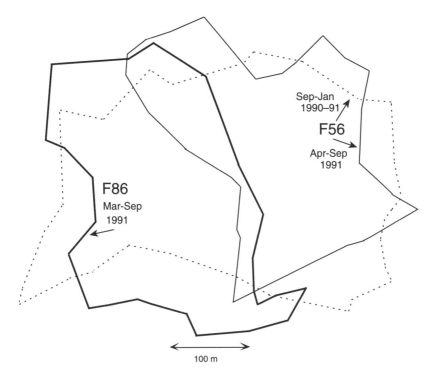

Fig. 9.5. Takeover in March 1991 of half of the territory of *T. longipes* female F56 by female F86, born on the study area in September 1990.

(a young animal, perhaps parous, that had lost her ear tag), and it is likely that F66 was still present but trap-shy. M287 had a gigantic territory 640 m long, and from its size I expect it included two females. Sadly, this male was accidentally killed when a trap door fell on him, and his home range is based on limited data.

Males and females of *Tupaia longipes* pairs that shared territories met briefly from time to time. Long periods (hours) were sometimes spent together at fruit trees, perhaps coincidentally. The pair at Poring (F132, M138) had a behavior never detected in any other *Tupaia* individual of any species: they shared three of the same nest sites (burrows), but they used them alternately, never together. The members of the other two radio-tagged *T. longipes* pairs did not even have burrows in the same parts of their mutual home range.

Plain treeshrews were thus territorial within sexes, with males either monogamous or perhaps overlapping a second female. Female territories were so large that, as with slender treeshrews, it is unlikely that males could have defended more than two female ranges (this would have been 17 ha at Danum in 1990). Because the territorial boundaries of males were closely congruent with those of females, and in all four cases where there were data males did not overlap neighboring females with resident partners, it is likely that males either defended a whole female territory or none of it. Nonetheless, from time to time males seemed to make sorties or prospecting trips abroad from their normal boundaries, as male mammals of many species are prone to do.

TUPAIA MONTANA   Montane treeshrews were studied for only a month, so the data are a snapshot of a few individuals. We radio-tagged four females and two males and tracked for only two five-day samples, when we tried to follow several animals simultaneously.

The pattern of home ranges shows within-sex territories, with neighbor boundaries tightly following each other's contours (fig. 9.6). The single pair for which there were good data for both animals had almost perfectly congruent male and female borders, implying a monogamous pair. The incomplete data for two other pairs are consistent with the same configuration, but there is not enough information to know if the males all had one-female territories.

The *T. montana* pair (F50, M143) spent two of the ten days almost always together, an hour on another day, and a few minutes on most of the remaining days. Because this species had small territories, their extensive foraging travels folded back and forth in a compact space, and pairs were bound to meet often even if they traveled randomly. This male and female nested so close together on some nights that from a distance we thought that they were in the same nest, but when we worked our way in through difficult, steep terrain, we found them in separate nests 4 m apart. These limited results seem to show that mountain treeshrews resemble plain treeshrews in their land tenure system, but pairs seemed to spend more time close to each other.

TUPAIA TANA   At Poring we trapped three adult female *T. tana*, one male and two subadults. We radio-tagged the male and two adult females. At Danum Valley we captured thirty-two individuals, of which about thirteen were subadults, and followed nine females and eight males by radiotelemetry. Because the array of animals was more complete, I describe the social organization only for Danum Valley, but the behavior of *T. tana* at Poring showed no salient differences.

As in every other *Tupaia* species followed, known resident adults of the same sex had nonoverlapping territories. Female territories in most

Social Organization

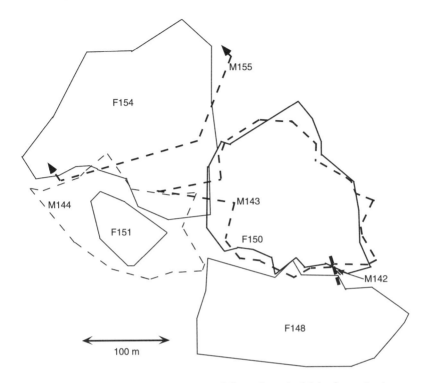

Fig. 9.6. Territories of all *T. montana* followed on the high plot at Poring. Note how the irregular range boundaries of females F148, F150, and F154 fit tightly together like pieces of a puzzle. Only one heterosexual pair is adequately represented (M143 and F150). Their ranges coincide almost perfectly.

cases had gaps between them. This might be due to insufficient data points, but that it is such a regular feature of at least eight territories implies active avoidance (figs. 9.7, 9.8). The small overlap of F78 and F54 was formed by just one location point for each individual (fig. 9.7). Most territories of males also had no overlap, but there were a few small zones with slight superposition. In some cases these seemed to be due to the location of baited trap sites: males M167 and M168 were trap-happy banana junkies that overlapped at two traps. This was also the case with M162, a subadult in the territory of F54 in early March and likely her son, who stayed in the same place as he matured. The other incidence of overlap among males was the rectangular piece of M77's range that is shown as a dotted line in figure 9.7. This whole section was the result of a single foray (Nov. 1990) in which he raced 600 m down and back to his starting point within about an hour. Since he never again was recorded in that area, although he lived until the end of the study (Sept.

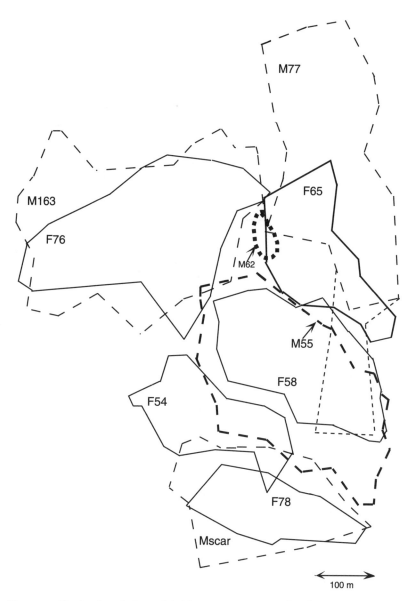

Fig 9.7. Territories of nine adult *T. tana* at Danum Valley from September to December 1990. The location of one subadult male (M62, not radio-tagged) is also shown. Solid lines = females; dashed lines = males. The line of small dashes below the range of M77 represents a single excursion of one hour, which was probably outside of his territory.

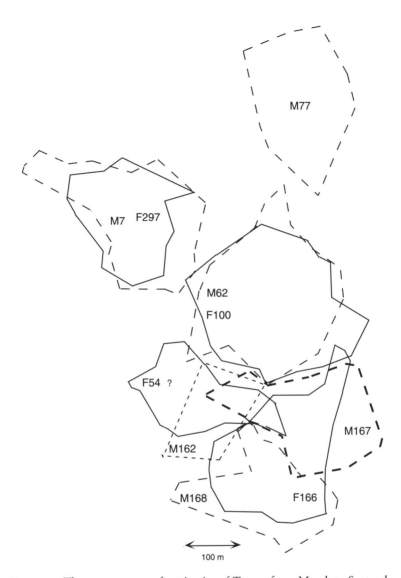

Fig. 9.8. The arrangement of territories of *T. tana* from March to September 1991. Solid lines = females; dashed lines = males. It is unknown whether female F54 was still present. Note the overlap of F166 with two males (M167 and M168). The intersection of several ranges on the east side of M162 is at some popular trap sites (see text). Because there were so many animals to follow, the data on individual treeshrews were collected in a time-staggered series, and because changes in territories happened at different times, not all the animals shown were present throughout all the months for which a figure is relevant. One female that had a recorded tenure of only a couple of months is not shown (although she had young and a territory), because it would give a false impression of overlap that did not exist.

1991), this zone was likely the result of a male sortie (to visit F58?) and outside of his territory.

Unlike other *Tupaia*, the territories of male and female large treeshrews coincided only approximately. Although most of the territory of a female was in six of eight cases in the home range of a single male, in every case the boundaries of males and females were not sharply concordant, and both males and females at least partially overlapped more than one member of the opposite sex. One male, M111 at Poring, probably overlapped two females, and M77 at Danum probably did. One female (F166) had most of the territories of two males in her territory (M167, M168; figs. 9.7, 9.8).

Both male and female large treeshrews thus had more than one possible overlapping mate, but most individuals were only recorded actually encountering the one animal of the opposite sex with whom they shared most of their territory. It was evident from following large treeshrews in the field that males were the active initiators of most encounters. As in the case of the lactating female F109 that we detailed at Poring (Emmons and Biun 1991), a male would travel to where a female was, usually early in the morning when she was foraging. He would spend from a few minutes to hours with her or trailing her (his signal following hers). The following pairs were registered as being together: F109 and M111 (on 7 days); F76 and M63 (2 days); F65 and M77 (5 days); F78 and Mscar (5 days); F297 and M7 (3 days). In addition, F100 was near M7 once, and F166 was near M162, but it was uncertain whether they met.

The most interesting behavior was that of F54 and Mscar, who were not principal overlapping mates (fig. 9.7). This male visited F54 on five days, and on three of these he also visited F78, sometimes more than once. As I followed him I was convinced that he would start looking for females, running without seeming to forage, first to where he would find F78 and then to the zones where he was likely to meet F54. These last places were on the extreme border of her home range, and she sometimes seemed to deliberately go and hang around there. F54 was once noted with another overlapping male, M55, under a fruit tree. Possibly these interactions with a neighbor male were due to the lack of her own mate. No known male overlapped most of her small territory, until her presumed son (M162) matured and stayed there. The way that territories fitted around the traplines precluded trapping *T. tana* on the southwest side, and it is possible but unlikely that F54 had a male that I did not catch. Because of the combined vicissitudes of trapping and timing of transmitter function, overlapping or neighboring animals could not always be simultaneously monitored during the same months.

Large treeshrew pairs thus seemed to be more loosely associated than those of other species, and their territory deployment implied that they might be able to choose between several mates by choosing which parts of their territories to use.

Just as with *T. gracilis* and *T. longipes,* there was a changeover in *T. tana* individuals during the year, with a stable set of adults present from September to December/January and a new set that replaced some of these and formed a new distribution of home ranges, from March to September (figs. 9.7, 9.8). In this last half of the year, when many new residents became established, territories were smaller and the density of animals increased (fig. 9.8, table 8.5). Two subadult females and one subadult male, all probably born on the study area, became resident adults. The fate of most of the animals they replaced—whether they died or were forced off their territories—was unknown. Male M77 "lost" the southern part of his territory and was living only at its northern end in August 1991, and I presume an unmarked male that was seen in the vicinity had taken over the vacated zone.

TERRITORIAL BEHAVIORS

During the entire study, I saw only four aggressive chases that could have been territorial fights. I saw treeshrews marking by rubbing parts of their bodies against the substrate only nine times: *T. minor* three times (two males, one female); *T. tana* four times (three males, one unknown sex); and slender and pentail treeshrews once each. Males dominated the tiny sample with six of seven records where sex was known or inferred.

DISCUSSION OF SOCIAL BEHAVIOR

SOCIAL SYSTEM

TUPAIA   The only other description of the social organization of wild treeshrews is in the excellent and detailed study of *T. glis* on Singapore by Kawamichi and Kawamichi (1979, 1982). Their dense study population of about thirty resident adults of each sex had a social system virtually identical to those I found in the *Tupaia* species in Sabah. Of sixteen male territories they recorded, fourteen were monogamous, with one included female, and two were polygynous, one with two females and one with three. One of the females of this latter triplet was associated with two males, with about half of her territory overlapping each (Kawamichi and Kawamichi 1979). The Kawamichis observed territorial chases 33 times and scent marking 110 times, with males marking significantly more than females. These frequent interactions were likely due to the dense population, while in Sabah the rarity of chases and marking behaviors can be ascribed to the much lower densities and rarefied populations of all species.

Based on their behavior in captivity, Martin (1968) correctly inferred that *Tupaia glis* (= *belangeri*) lived in groups no larger than a family unit of two parents and their young. He surmised that the male and female were strongly pair-bonded, because captive pairs slept together in the same nest box, rested in contact, and marked each other. None of these behaviors was seen in the species studied in Sabah; but nonetheless, the concordance of territorial boundaries, and the frequent encounters between co-owners of territories, implies stable bonds between males and their one or more females, though not strict monogamy. Because the *T. glis* studied by Kawamichi and Kawamichi (1979) exhibited social organization apparently identical to that of *T. longipes, T. minor, T. montana,* and *T. tana* in Sabah, it is likely that the laboratory-bred stock of *T. glis* (of unknown origin) studied by Martin (1968) was not originally fundamentally different from other *Tupaia* and that the nest-sharing and other cohesive behaviors were artifacts of the close conditions of captivity.

The Kawamichis used the term "solitary ranging pair" for this social system of solitary animals that are monogamous on a common territory that each sex defends against its own gender. This has also been called Type I or "facultative" monogamy (Kleiman 1977). The Kawamichis cited a few other species with this system, including some nocturnal prosimians (Bearder 1987) and pikas (*Ochotona* spp.; Smith and Ivins 1984), but it is actually found in many mammals, including elephant shrews (Rathbun 1979), maned wolves (Dietz 1984), agoutis (Dubost 1988), and others (Kleiman 1977). There is no fundamental difference between solitary territories where a male overlaps one female and those where he may overlap one, two, or more, as in *Tupaia* species, except that as the system becomes more skewed (more females sequestered by each male), there are other developments such as sexual dimorphism and skewed reproductive success. This last may be one of the most common social organizations among "solitary" territorial mammals, such as galagos (Charles-Dominique 1977), ocelots (Emmons 1988), spiny rats (Emmons 1982), and probably hundreds of rodents. Little phylogenetic

Social Organization   163

significance can be attached to the social organization of *Tupaia*, because it is a simple system that occurs again and again in many orders throughout the Mammalia. It is more interesting as a reflection of ecological factors.

Obligate, as opposed to facultative, monogamy in vertebrates such as birds (where it is the dominant system) and mammals such as foxes and small primates is generally agreed to be the result of increased offspring survivorship if both parents contribute to their provisioning or care (Kleiman 1977). I believe that in treeshrews and other species with facultative monogamy, where the father has no direct role in caring for the young, the defense of a territory where young remain until they reach adult size is a large indirect contribution to their nutrition. This may be as important to their survival as more direct provisioning in obligate pairs. If a male is able to defend two females, then each of their territories has only half a male feeding on it, which can only be to their advantage.

Arboreal, terrestrial, lowland, and montane species of *Tupaia* all had versions of the same social organization, but especially in *T. gracilis* and *T. tana* variations in the number of female territories on a male territory, or vice versa, and the deviations from exact concordance of male-female boundaries give both sexes possible access to several mates within the system of monogamous pair territories. It would be interesting to test the paternity of *Tupaia* young, to see how monogamous the social system really is.

PTILOCERCUS   From all meager evidence, like *Tupaia*, pentail treeshrews seem to be monogamous but with a completely different social organization. Unlike *Tupaia*, pentails have a den-centered existence. An adult pair and their offspring always used a single nest site together and radiated their foraging centrifugally from it. If the pair lives together permanently with their young, they would seem to have cohesive, obligate or Type II monogamy. It is yet unproven that the male stays permanently with the female; and the length of tenure of the young with their parents is unknown. Pentails foraged alone, perhaps in partially exclusive foraging domains within the area used by the group. Group life in pentails therefore is not likely to be related to shared foraging information or security from predators while active, because they avoid each other while foraging. This pentail social organization resembles that described for a tiny emballonurid bat, *Rhynchonycteris naso* (Bradbury and Vehrencamp 1976), in which adult members of a colony use individual solitary foraging beats, juveniles forage near the colony, and an adult male uses the extreme periphery, outside the core area, which he defends. The bats are territorial, and I suspect pentails to be, but there is no evidence for it at this time. This type of system seems likely to arise when a pair or a group is needed to effectively defend a large foraging space or a sleeping site.

A single den tree formed the hub of pentail social life, and perhaps some special safety or other feature of the den site makes it a rare commodity in which a pentail family forms a nucleus. Because pentails are torpid and defenseless by day (Whittow and Gould 1976; Gould 1978, pers. obs.), they may require unusually well protected sites, such as hollow, high canopy branches with small entrance holes (perhaps from within the tree) that predators such as monkeys, martens, weasels, and snakes cannot enter or dismantle. In captivity *P. lowii* has the odd behavior of liking to sleep in glass jars (Lim 1967; Gould 1978). Gould (1978: 6) reported, "Despite an abundance of snug and quiet nest boxes as many as three pentail shrews crowded themselves into a single jar." This mystifying behavior (the jars were clear, transparent glass) may indicate a preference for tight, smooth-walled narrow tubes such as branch hollows. Grouping also may have physiological advantages. If the diurnal torpor is an energy-saving mechanism (implying that pentails have an energy problem), then group huddling while inactive, as pentails do in captivity, may conserve heat and energy (Whittow and Gould 1976; Gould 1978).

TERRITORY FORMATION AND FAILURE

The species of *Tupaia* differed in the cohesiveness of pairs that shared territories, as defined by the amount of recorded interaction. *Tupaia minor* and *T. montana* pairs spent the most time together, including some complete days of activity. These two species had the smallest home ranges (table 7.2) and were thus the most likely to be near each other by chance, but lesser treeshrew pairs often traveled together while insect foraging. The other three *Tupaia* had only brief contacts, sometimes daily, which consisted of male-initiated visits to females, within solitary foraging itineraries.

An intriguing behavioral question is how pairs of such solo-foraging and independent individuals as *T. longipes* make their extended boundaries so strongly concordant over as much as 1,000 m of borders. Kawamichi and Kawamichi (1979) suggested that pair members use scent marks to adjust their boundaries to those of their partners, a likely possibility. If true, I surmise that only the male fits his territory over the range

Social Organization   165

of "his" female(s), whose own territory is determined only by the landscape and the borders of neighboring females. A male's goal would in this case be to exclude the neighboring males from any overlap with his mate. The sloppy territorial boundaries of *T. tana* with respect to individuals of the opposite sex seem to imply either weaker pair bonding, tenure of males too short to establish tight pair borders, or inability of males to defend all areas where females encroach other males. Another possibility is of a different, more polyandrous/polygynous social system, in which the male and female territory maps arise independently of each other and are controlled only by within-sex interactions, so territories are placed wherever an individual can muscle in among members of its own sex. Too much coordination of pair territories is present for this last to seem likely. For example, in 1990 *T. tana* M62 (not then radio-collared) was present in a small area between other males, but when M55 died and F58 and F65 both vanished, he expanded into a territory of an entirely different shape from those of his predecessors, and coincidentally F100, daughter of F58, acquired a territory that tightly overlapped his but included half of her mother's old area and half of F65's (figs. 9.7, 9.8). Together these two set up a new territory with new boundaries. The similarity of many large treeshrew male and female superimposed areas implies joint coordination between pairs (fig. 9.8). A critical feature of *T. tana* male territories is that they are larger than female territories but not usually large enough to include two females. Males thus encroach slightly on neighboring females, and at odd points, females on male neighbors. The only two likely courtship interactions that I saw involved only primary resident pairs (F109 and M111, F65 and M77). Males also have larger apparent ranges than females because males make sorties into neighboring zones, probably trespassing to investigate neighboring females and/or the presence of other males.

In *T. gracilis* and *T. longipes* I witnessed the formation of new territories by females that slowly pushed resident females aside and took over all or part of their space within the territories of resident males (F173, F86). This gradual encroachment is distinct from the all-or-none results of contests between mammals that fight physically for dominance. I suspect that these female-female interactions took place without much influence by the males: the male slender treeshrew visited the areas of each of the two females almost daily, and it seemed that he acquired two females by their spatial rearrangement, without changing his own. This was probably also the situation with *T. longipes* M64 on the territories of F56 and F86, but his boundaries during their interactions were unknown. The spatial arrangement of *Tupaia* territories is therefore the result of complex interactions of each sex with both the animals of the same sex that surround it on all sides and the territorial boundaries of their own mates. In at least some species, the same-sex interactions sometimes result in polygynous or polyandrous deployment of opposite-sex territories, although the norm for all *Tupaia* species studied in the field so far appears to be monogamous pairs that largely overlap the same ground.

Despite the presence of clear territorial boundaries, the territoriality of *Tupaia* species has some flexibility. The Kawamichis saw the breakdown of territoriality at a large fruit tree:

> From early February until mid-March, one large fig tree (*Ficus dubia*) provided an ample supply of ripe fruits. The sweet smell was perceived by us 250 m from the tree. Besides a pair . . . whose territories included the fig tree, all nine adjacent residents and seven other residents from the surrounding area came to the tree. . . . In addition, nine non-residents including six vagrants were also counted there. . . . The owners of the tree chased the visitors of the same sex. By mid-March the figs were eaten or rotten, and the residents subsequently confined themselves to their previous ranges. (1979: 392)

At Danum Valley bait (trap) sites had a similar effect of drawing some *Tupaia* into the territories of others. Prebaiting probably enhanced this behavior. The ranges of one pair of radio-tagged *T. tana* also seemed to be extended for a few days by some long daily journeys to a fruiting *Polyalthia sumatrana* that was signaled by noisy flocks of hornbills and other birds. The treeshrews could have detected the bird activity at this tree from afar. Possibly territory owners do not or cannot invest much energy in defending large fruit sources that vastly exceed their own needs.

ECOLOGICAL ASPECTS OF TERRITORIALITY

The solitary foraging, long daily active period, and large daily movements of treeshrews show that their food resources are scattered, small, and hard to find. Because they use almost all of the time available in a day for foraging activity, there is not much leeway for more. Defense of feeding territories may sequester the food supply and allow treeshrews to feed within a smaller time and space than they would if conspecifics competed on the same feeding grounds. Treeshrew territoriality is very likely the simple defense of adequate food resources for reproduction coupled with the sequestering of his mate by a male. That territories of *T. glis* in Singapore were one-tenth the size of those of the similar *T. longipes* in Sabah (Kawamichi and Kawamichi 1979), but the social system was identical

Social Organization

to that in Sabah under tenfold lower population density, is evidence that the basic system of pair territories is robust to opposite resource extremes. In the next chapter we shall see that the *T. glis* in Singapore and peninsular Malaysia appear to have compressed their territories (or the Sabah *Tupaia* expanded theirs) such that they achieve similar resource levels as measured by reproductive output.

Unlike most other nonvolant mammals, treeshrews have no sexual dimorphism in body size (see table 2.2). This is consistent with a social system in which long-term monogamous pairs are the norm, effective sex ratio is not highly skewed, and territorial possession is not determined by dramatic physical dominance or fights (Kleiman 1977). The slow territorial displacements we registered do not suggest that territories are contested by deterministic battles in which the strongest or largest animal wins during a single contest. The energetically difficult life of treeshrews may give them little time for activities other than foraging, and absence of sexual dimorphism may also reflect strong energetic constraints on body size. A pair of solitary foragers on a common territory may be the most efficient use of both space and time for cryptic animals for whom sociality carries no feeding or predator escape advantages.

WHAT SOME OTHER SHREWS DO

The social organizations of true shrews and elephant shrews illustrate the kinds of patterns found in other primitive mammals and give a perspective on those of treeshrews. The true shrews are the most energy limited of all mammals, and energetics overshadows their lives, but the two subfamilies, Soricinae (including most New World and many Old World genera, such as *Sorex, Cryptotis,* and *Blarina*) and Crocidurinae (including many Asian and tropical genera, such as *Crocidura* and *Suncus*), seem to have different life history characteristics, or strategies (reviewed in Churchfield 1990). *Sorex* species, where known, form exclusive territories as subadults or young adults, with no overlap between sexes or individuals during the winter, but in the summer breeding season males start to wander over female ranges, and offspring overlap with adults. Males seem to have no positive interactions with females or young apart from copulation. In the larger *Blarina brevicauda* territoriality in winter is exclusive between all individuals, but in the summer it is exclusive within the sexes, and males and females overlap. *Cryptotis parva* are gregarious and live in small colonies where adults share the same nest and home range. In several Crocidurines (*Suncus* and *Crocidura* species) there is apparent obligate monogamy, wherein the male shares the nest with the female and young, helps to build it, and shows some parental care behaviors toward his young. *Crocidura russula* and *Cryptotis parva* are not territorial. Elephant shrews of several species show strict, within-sex territoriality, wherein monogamous pairs share contiguous territories (Rathbun 1979), but members of pairs forage alone and interact only occasionally. If a male disappears, a neighboring male can expand to cover two females, but this situation is temporary.

These other small insectivorous mammals therefore have an array of social structures some of which are the same as those of tupaiids. Even such tiny, short-lived mammals as shrews express a wide spectrum of social organization, including solitary, monogamous, and group living species, and spatial organization that ranges from extensive overlap to territoriality between or within sexes. Advanced cognitive powers are not needed for expression of any of the mammalian social systems, which show great variability within families and even within genera. *Tupaia* species have a social system like that of some prosimians, but so do short-tailed shrews, agoutis, elephant shrews, and spiny rats. The spatial organization of mammals is plastic and changes to accommodate changing ecological conditions.

CHAPTER 10
Life History

Short-lived mammals, for example, must always gamble somewhat with their reproduction. They must reproduce as continuously as possible to counterbalance their short life expectancy. Longer-lived mammals can afford to be more selective. All other things being equal then, short-lived mammals tend to be more opportunistic and less seasonal in their reproductive effort than do long-lived mammals.
F. H. Bronson, *Mammalian Reproductive Biology*

REPRODUCTION

THE ABSENTEE SYSTEM AND PARENTAL CARE

One of the most remarkable features of *Tupaia* biology is the "absentee" maternal care system. Martin (1968) first described this behavior in captive *T. glis belangeri* and then in captive *T. minor* and *T. tana* (D'-Souza and Martin 1974). The basic features of the system Martin (1968) described are (1) the mother gives birth to her young in a nest that she never shares with them, where they stay until weaning; (2) the mother visits the young to nurse them only once every other day for less than five minutes; and (3) the mother does not show any of the care behaviors most mammal mothers have, such as retrieval of young in distress or grooming or cleaning the young or nest. *Tupaia* young are altricial: they are born as tiny, hairless pinkies with closed eyes and ears. For the absentee system to work successfully, which it does, an extensive suite of physiological and behavioral adaptations are needed. The nursing behavior of *T. tana* is described by Miles Roberts, who placed a video camera inside nest boxes with young at the National Zoo:

> [The babies] have slept deeply and undisturbed for more than 40 hours, but during the last six hours they have awakened from their deep sleep every 15 to 30 minutes to groom themselves and to remove any traces of urine, droppings, and other material that might leave a telltale odor for a passing predator. . . . [T]he babies sense [the mother's] presence through vibrations caused by her movement on the outside of the nesting cavity. In response, they push

themselves up on wobbly forelegs and thrust their heads up as high as they can, probing the air with their tiny snouts.... Waving their heads in the air like leeches searching for warm-blooded prey, the babies suddenly pick up her odor.

The mother enters the nest chamber slowly and picks up a leaf.... Quickly, she approaches the babies in the dark.... Without as much as a nuzzle or a lick of greeting, she rears up on her hind legs and extends her forelegs to expose her abdomen to the anxious babies.

The babies frantically nose her abdomen.... Within seconds, each makes contact and nurses at one of her four nipples with what can only be described as hysteria. After a few seconds, each baby switches to another nipple, nurses a few seconds, then switches again to another. Slow at first, the switching tempo quickly increases as the mammary glands empty....

After 60 seconds of nursing, in which they consume nearly one-third of their body weight in milk, the babies are bloated but continue to nurse vigorously. Now, the mother becomes restless. Suddenly, with no warning, without as much as a departing look, she simply vanishes. The babies, engorged and seemingly intoxicated with milk, collapse onto one another and laboriously reconfigure themselves into the huddled ball that will best conserve their body heat. Within seconds, they enjoy the sleep of the dead.

... Except for nursing the young for 90 seconds every other day for the next four weeks—a sum total of about 25 minutes—[the mother's] investment in the young is virtually complete the moment they are born. (Roberts 1993: 6–7)

On this regime captive young *Tupaia* grow rapidly, almost doubling in weight weekly, until they are weaned and leave the nest at about 25 to 33 days old (D'Souza and Martin 1974; Martin 1968; Roberts pers. com.). In captive colonies weaning is usually abrupt, on the day that the young emerge from the nest box, and from then on the mother does not pay any attention to her offspring. Pentail treeshrews have not been reported to breed in captivity, and their maternal behavior is undescribed.

Elegantly and repeatedly documented in captivity, this curious absentee system still had not been shown to exist in the wild when I started my fieldwork. D'Souza (1972) had reported a possible tree hole nest with young of *T. minor*, but there were two puzzling inconsistencies in the observations. First, the hole was visited once each day, rather than every other day, and quite late (0900–0915 h); on one day it was visited twice, with an evening visit at 1800 h. Second, D'Souza stated that "characteristic vocalizations of the young" were heard during these visits. Roberts (pers. com.), who recorded with video cameras inside the nests of *T. minor* and *T. tana*, never heard young vocalize under any circumstances unless removed from the nest by a human, in which case they emitted only a faint aggressive sound. Paramount among my goals in launching a field study of treeshrews was to observe the absentee maternal care system in the wild.

TUPAIA TANA   Despite much directed effort to find nests with young (chapter appendix), the single "nursery" nest found during the project was by serendipity. On 7 April 1989, at the beginning of the study at Poring, we captured a lactating *T. tana* (F109) and radio-collared her. Lactating treeshrews captured early in the morning before feeding their young had a sheet of milk covering the abdomen below the skin, whereas those captured following nursing had no milk visible at all. I started a five-day continuous sample to follow F109 on 17 April (we had followed other collared treeshrews in the meantime), and on 18 April at 0706 h, after she had been active for about an hour, I saw her come out of a tree hole that was in plain view of the trail (see fig. 6.6). I instantly suspected that this was the place she had put her young and carefully followed her for four more days. She visited the tree again on 20 April, and I again saw her jump down from it on 22 April, although she had not been there on 17, 19, or 21 April. I had observed unambiguous alternate day visits! Alim used his forester's skill to build a small platform completely screened with palm fronds, 2 m up in a tree a few meters from the nest hole. From there we could watch and videotape F109's visits. In the predawn dark one of us climbed into the blind and waited silently, listening for F109's radio signal through an earphone. Six times we saw her come, but we also watched on some "alternate" days, to verify that she did not approach. We intensively monitored this female through twenty-three days of lactation when young were in the nest and then monitored her for an additional twenty-two days after emergence of the young. We logged fifteen dawn to dusk days of radio-tracking F109 and many partial days. Our detailed observations were published earlier (Emmons and Biun 1991).

On 26 April, just after a nursing visit, we climbed to the nest to find out what it contained. We discovered two young, which we extracted briefly to weigh, measure, and mark with ear tags (fig. 10.1). The young were in a tree hollow with a large entrance 4 m high (fig. 6.6). Within the cavity they were inside a large clean nest of overlapped leaves lined with teased fibers. From the blind we watched six nursing visits to the tree hole. The mother indeed visited her nestlings only once every other day, and she spent an average of 2.74 min within the hollow (N = 5 timed visits; range 1.8–4.48 min). The nursing visits were always early in the morning, between 0615 h and 0734 h. She was videotaped on one visit,

Fig. 10.1. Nestling *T. tana* on 26 April 1989. Note immobility (it does not move or try to escape).

and on the tape we saw that she carried a single leaf in her mouth into the hole but left without it.

A notable feature of F109's behavior when she came to nurse her young was that she often varied her pathways, coming and going by different routes. Sometimes she jumped from tree to tree on slender understory saplings, without touching the ground anywhere near the nest tree. She usually did not run up or down the nest tree but jumped to it at the level of the entrance from a neighboring treelet (see fig. 6.6). The radio signal indicated that she often hesitated nearby before running to the nest. Before exiting, she would poke her head out of the hole and scan the surroundings. We never heard any vocalizations or deliberate sounds made by mother or young, and visits were as discreet as possible. At the hour of most visits, the light intensity was still crepuscular, which made it somewhat difficult to see the treeshrew or to get sharp video images. These cautious behaviors should hide the nest from predators that could use the mother's visible behavior or odor trail to find the young.

The pattern of alternate day visits by the mother *T. tana* only varied once, when on the day before the young emerged the female fed them on an "off" day, so that we assume she came on three consecutive days, the two "normal" ones and the one in between (we did not follow her

Life History

the fourth day previously, so we cannot state she did not visit then). Martin (1968) had noted that in captivity females come to nurse on an extra day on the day before the young emerge, so we anticipated the event. The young emerged from the nursery nest on 10 May, at a minimum of thirty-four days old.

On the day that the young emerged, Alim observed the following behavior from the blind. The female arrived at 0643 h, carrying something in her mouth, and went into the nest. Two minutes later she exited slowly and stopped to wait on the branch of a sapling. The two young came out of the hole and shakily climbed around on the side of the tree. When the young reached the ground, wobbly and uncertain, the mother went away for eight minutes and returned with a food item that she gave to them and they ate. She also licked them. She led them away from the nest site by repeated calling, after a period of "training" them to follow her by calling, moving a little away, and waiting for them to approach, then moving away again and calling. The mother then spent the whole day and the first night with her young. They slept on the ground surface in a thicket under the shelter of a log. The second day, she led them to another nest site in a dense treefall, but that night she went back to her own previous nest alone and thereafter did not share nest sites with her young. The mother spent much time with her youngsters after their emergence, including the whole of the first two days (apart from racing once down to the stream and back, presumably to drink). She visited them every day in the morning and often spent many hours with them. She continued to spend time with her young to twenty-two days after emergence, when we had to discontinue tracking. Unfortunately, we did not see what she was doing when with her young, as we glimpsed them only fleetingly; but once a juvenile was seen foraging by itself when she was present. She visited the place where her young were early each morning between 0624 and 0717 h, but otherwise her behavior was unpredictable. She spent the whole day with them on 15 and 18 May and more than three hours on 31 May but only seven minutes on 30 May and eleven minutes on 19 May. Soon after they emerged the young foraged on their own. On 13 May they were seen foraging twice, once biting apart dead branches. The early morning visits suggest that she may have continued to nurse the young after emergence.

THE STRANGE CASE OF THE MISSING MALE   At Poring an enigmatic social arrangement of *T. tana* took place when the female had young. We only observed this once, and we previously described it (Emmons and

TABLE 10.1. The behavior of *T. tana* male M111 at Poring after the birth of F109's young on their shared territory (from Emmons and Biun 1991).

Date	Young	Interactions between Male and Female
7 April	In nest	F109 first captured, lactating
17 April	In nest, not nursed	Together 0727–1105 h and 1338–1420 h
18 April	In nest, nursed	Not together all day
19 April	In nest, not nursed	Together 0753–0941 h
20 April	In nest, nursed	Not together all day
21 April	In nest, not nursed	Together 0642–0710 h
22 April	In nest, nursed	Together 0635–0710 h, loud, continuous chattering
25 April	In nest, not nursed	Not together all day
27 April	In nest, not nursed	Not together all day
28 April	In nest, not nursed	Together 1103–1400 h
29 April	In nest, not nursed	Together 0950–1006 h
7 May	In nest, not nursed	Male nesting at night on female's range
10 May	Young emerge	
9–13 May		Male apparently dissappears from area
14 May		Male found once on F109's territory
15–29 May		Male not on female's territory
30 May		Male found nesting far across ravine outside F109 range
5 June		Male nesting far across ravine
27 June		Radio of male found cast on F109's range
15–16 August		Male captured three times on F109's range

NOTE: The male was recognized by his radio signal, which was searched for often, especially after it "disappeared."

Biun 1991), but it is interesting enough to recount again, for it may be important. When the nurslings were in the nursery nest, the male spent much time with the mother (table 10.1), but initially he did so only on days when she did not nurse. She may have been coming into estrus, as postpartum estrus during lactation was reported by Martin (1968). We often monitored the radios of both male and female (six days a week we tried to find every treeshrew radio every night, to pinpoint nest sites, and whenever we followed a *tupai* we periodically searched for all other radios). Initially, the male M111 was always somewhere on the range he shared with F109, until suddenly, on about the day the young emerged, he abruptly vanished from her territory. Apart from a single return, he was gone from it for a month or more. His signal suddenly reappeared,

Life History

and we found his slipped collar on the ground in the middle of F109's range on 27 June. We do not know exactly when he returned, because we stopped listening for him on her range as we believed that he had permanently decamped to a zone outside our study area where we could only rarely pick up his faint signal (and only at night when he was up in an arboreal nest, where we went to find him once). We caught him on F109's range three times when we trapped again in August, and it seems that he had returned to live there. Did F109 drive him away when her vulnerable youngsters emerged? Or did he move in with a neighboring female who was a better breeding prospect? I do not know, but I once saw F109 violently drive another *T. tana* away from the vicinity of the nursery tree.

OTHER TUPAIA SPECIES   Field evidence for absentee maternal care and alternate-day nursing in other *Tupaia* species was circumstantial. (1) The nest of a lactating *T. montana* was inspected and found to contain no young. (2) Lactating females of *T. gracilis, T. longipes,* and *T. tana* each used several nests within a few days, so the young could not have been in them with the mother (female *Tupaia* have no behaviors of retrieving or carrying young [Martin 1968]). (3) While they were followed continuously, lactating females generally showed alternate-day patterns of visiting certain areas in early morning and did not show patterns of sites visited at dawn daily or more than once daily. But because nursing visits are so short, and animals in general tend to repeat their favorite travel routes on their home ranges, it was always difficult to interpret their movements unambiguously. I tried to locate nursery nests many times, with different individuals and species, but never succeeded again.

PTILOCERCUS   The adult female pentail (F181) was lactating when captured on 22 August 1991. On 28 September I saw a very small young emerge from the nest tree at nightfall within 10 min of when F181 emerged, as well as another large adult and a smaller, probably subadult pentail. The following day, when at least two others emerged with the youngster, the female and another stayed near the nest tree and baby and repeatedly made alarm calls (probably because of my presence). On 4 May 1991 I saw two large pentails emerge, followed 10 min later by two apparently slightly smaller pentails (presumed subadults). On 9 May several of us saw a minuscule pentail emerge from a hole at the nest tree, with at least two others. Again an excited adult remained near the tree, calling at us. Baby pentails therefore shared the maternal nest tree from at least what seemed to be the age of first emergence.

TABLE 10.2. Litter sizes of the *Tupaia* study species, from field collections only.

Species	N	Embryo No. 1	Embryo No. 2	Reference
T. minor	4		4	Davis 1962
	4		4	Harrison 1955 (West Malaysia)
	1	1		This study
T. gracilis	3		3	This study
T. longipes	2		2	Davis 1962
	3		3	UKMS[a]
	3	1	2	This study
T. montana	10	1	9	Kobayashi, Maeda, and Harada 1980
	1		1	UKMS
T. tana	3		3	Davis 1962
	1		1	Kobayshi, Maeda, and Harada 1980
	1		1	UKMS
	1		1	Lim 1965
	4		4	This study

NOTE: Data from this study are based on palpations of embryos in living animals, and from one litter in the nest.

[a] Specimens in Universiti Kebangsaan Malaysia, Kampus Sabah, Museum.

I cannot exclude the possibility that the young pentails were in a different, absentee nest before emergence and that the mother then led them to her own nest after emergence (as the *T. tana* mother led her young), but because of the communal nesting behavior of this species, this scenario seems much less probable than that the young shared the maternal nest from birth (see chaps. 6, 9). However, the lactating female pentail did visit one area on seven of nine nights she was followed (all prior to 28 September), often about 30 min after she became active. This was a vine-choked streamside thicket, where I could not see her. She may have gone there to drink, as lactation would increase her fluid needs. She also visited another area on all nine nights, but at different hours, and she regularly repeated several travel routes. Thus her movements during early lactation did not rule out absentee maternal care, although I believe such a system to be unlikely.

LITTER SIZE

Embryo numbers or litter sizes of the *Tupaia* study species in the wild seem to be fixed at two, with the occasional single young (table 10.2).

Life History

Reports of litter size for West Malaysian *T. glis* are likewise constant at two (Harrison 1955; Langham 1982). For various subspecies of *T. glis*, Lyon (1913) reported two litters of three (*T. g. ferruginea*) in addition to eight litters of two and one singleton. He also noted a *T. chinensis* with a litter of four and a *T. nicobarica* with a singleton. These last are of interest because *T. chinensis* is one of the few species with three instead of two pairs of mammae, and *T. nicobarica* is one of the few forms with only one pair (Lyon 1913). There do not seem to be any published records of litter size for *Ptilocercus lowii*, but we saw two different, single, juveniles emerge from the den, while earlier in the year two like-aged subadults lived there, evidence for litters of one or two.

BIRTH SEASONALITY

Juveniles and subadults appeared erratically during the monthly trapping at Danum Valley. Most young were not likely to encounter a trap until they were traveling some distance, at one or two months post-emergence (about three months old), so young were quite large at first capture, doubtless because of the combined effects of the long interval between trapping periods, the restricted area used by young, and the narrow region of the traplines. Samples were small for all species, but the combination of dates when young were captured and when females were pregnant or lactating gives a fairly complete history of the reproductive activity of *T. gracilis*, *T. longipes*, and *T. tana* on the study area at Danum Valley during the thirteen months of the study (table 10.3). Evidence for the breeding patterns of *T. minor* and *P. lowii* was more spotty.

All *Tupaia* species at Danum Valley showed similar reproductive seasonality. There was a broad breeding season from August to November and a second one from March to May (figs. 10.2, 10.3). Between each period of increased breeding, there was an inactive period of two months. We saw pentail treeshrew juveniles in May and September, and we caught a subadult in November (probably born in Sept.–Oct.), evidence for the same general pattern. There was a slight seasonal difference in breeding between *T. tana* and *T. longipes* at Danum (figs. 10.2, 10.3). In 1991 the large breeding peak was a month earlier (Aug.–Sept.) in *T. tana* than in *T. longipes* (Sept.–Oct.), which species also virtually lacked the second breeding peak in late March.

In captivity treeshrews can breed continuously, with females becoming pregnant in a postpartum estrus and birthing every forty-five to fifty

TABLE 10.3. Reproductive histories, from monthly trapping, of resident female treeshrews at Danum Valley in 1990–1991.

Female	Sept.	Oct.	Nov.	Dec.	Jan.	Feb.	Mar. 1	Mar. 2	Apr.	May	June	July	Aug.	Sept.	Total
Ptilocercus lowii															
Pl 181			Y(2)		No Data					Y			L	Y1	3
T. gracilis															
Tg 67	Y1						P	LY1		Y1?					3
Tg 173								P		Y2					2
T. longipes															
Tl 56	Y2	PY2	L	Y1										Y1	4 (3)
Tl 66		Y2	L	Y1							P		Y1[a]		3
Tl X[b]		Y1+2								Y1					3
T. tana															
Tt 76	LY1	P		Y1				Y1							3
Tt 54	P Y3		PL				Y2+1								4
Tt 58		L		Y1											1
Tt 78	Y2														1
Tt 176								L							1
Tt 297														Y	1

NOTE: Y = young captured on territory; P = pregnant; L = lactating; Yn+n = young of two sequential litters. Total = probable number of litters during months monitored, (3) = number in the calendar year, when animal was monitored longer. Few females were present all year, and some were rarely captured. Female *T. tana* at the end of the list were new territory holders that replaced previous females. Litter numbers are estimated from the reproductive condition of captured females and the ages of young trapped on their territories. Youngsters from two litters were sometimes present simultaneously.

[a] F66 was neither recaptured nor replaced by another after December. She may have been present but trap-shy; this litter was on her territory.
[b] No female *T. longipes* was trapped on the NW corner of the study area, but one was present, as young appeared.

Life History

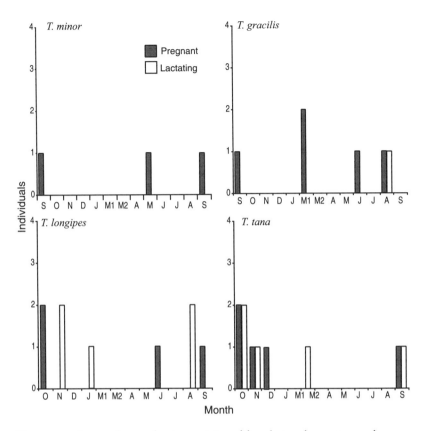

Fig. 10.2. Seasonal reproductive activity of female treeshrews trapped at Danum Valley.

days (Martin 1968). A pair of *T. glis* kept by Martin produced nine successive litters in a calendar year. In colonies at the National Zoo, lesser and large treeshrews also bred continuously, without evidence of seasonality (M. Roberts pers. com.).

In sharp contrast to the laboratory scenario, reports from field captures and collections of wild treeshrews all show a low overall pregnancy rate. In data from Sabah that Wade (1958) compiled from the April to August collections of Davis, only ten of forty-five adult females were pregnant (22%; three species): none in April or May, two in June, seven in July, and one in August. In West Malaysia Harrison (1955) found a crude pregnancy rate of only 9 percent for both *T. glis* (N = 53) and *T. minor* (N = 44). On Singapore Kawamichi and Kawamichi (1979) found no breeding in *T. glis* during their study from October to December. In

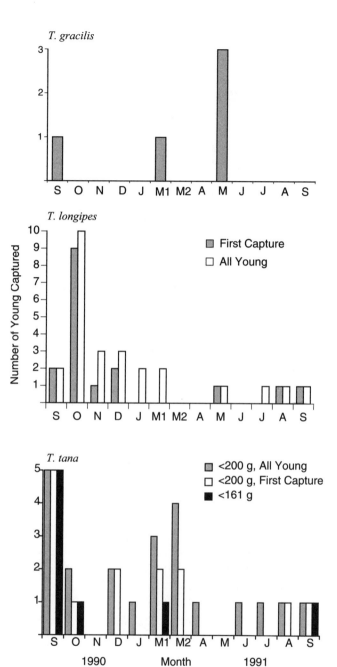

Fig. 10.3. Seasonality of the appearance of young in monthly trapping on the study area at Danum Valley. For *T. tana* only, young are separated into age groups by weight. All immature treeshrews that are nonparous or do not have adult dentition are classed as young.

Life History

Langham's (1982) three-year intensive trapping study in West Malaysia, with 256 monthly records of female *T. glis*, the number of reproductive females per month varied from 0 to 43 percent. He found a single, extended breeding season, from January to August; but most reproduction was from February to June, with few (<13%) breeding in January or July–August. No females were reproductive in September, October, or December, and only two in November.

Field studies thus show that wild *Tupaia* species have clear breeding seasonality, especially a lack of breeding in certain months. Moreover, although the areas from which data were drawn are at nearly the same latitude, the reproductive season differs between East and West Malaysia. My data agree with the findings of Wade (1958) for the same region of Sabah. Likewise, more than eighty squirrels in the same collections showed no reproduction in April or May, and only a small amount in June–August (Wade 1958).

REPRODUCTIVE OUTPUT OF INDIVIDUAL FEMALES

For some resident females at Danum Valley, I can infer the timing of sequential litters from the combination of their reproductive condition when trapped or the captures of young on their telemetry-defined territories. Most captured young can be assigned to a specific female. Two young of equivalent weight on the same territory I assume to be littermates. However, there are some ambiguous cases of young caught on two neighboring territories. Juveniles may not always be aware of territorial boundaries.

The data for individual females show that during each of the broad breeding seasons, some females bred twice (table 10.3). In particular, during the great breeding peak of August–November 1990, most *T. longipes* and *T. tana* had two litters in rapid succession, and one female of each of these species seemed to have three litters in this season (F56 and F54). In the March–June season only the two *T. gracilis* females had two successive litters. During the calendar year of the study at Danum, the individual females, or the successive females on a given territory, produced three to four litters of young each, usually in a pattern of two successive litters in one season and one in the other. Because few females were followed throughout the entire period, better accuracy cannot be achieved from this data set. The females captured in the five months at Poring indicated a similar reproductive seasonality for the lowland population (table 10.4), with breeding in March–April and beginning again

TABLE 10.4. Reproductive histories of female treeshrews at Poring Hotsprings in 1989.

Female	Apr.	May	June	July	Aug.
T. minor					
Tm 112	Y				
Tm 119					L
T. gracilis					
Tg 105	PY2				N
T. longipes					
Tl 132		L			P
Tl 133		N			N
Tl 155[a]			P		
T. tana					
Tt B	L	Y			
Tt 109	LY2				N
Tt 106	N				N
T. montana					
Tmo 141[a]			N		
Tmo 148[a]			P	L	
Tmo 150[a]				P	
Tmo 151[a]				P	
Tmo 154[a]				E	

NOTE: Abbreviations as in table 10.3; E = estrus; N = parous, nonreproductive, territorial female.
[a] At Langanan study site (900–1,000 m) trapped only in late June and July.

in August; but *T. montana* and a *T. longipes* captured in montane habitat at 950 m were apparently breeding on a different cycle, in late June and July.

Because litter size is nearly fixed at two, a female of *T. gracilis, T. longipes,* or *T. tana* would have produced six to eight young in 1990–91. Combined with the data on population density of females (see chap. 8), for six young this translates to an annual production of about 1.5 young/ha for *T. tana* and about 0.6 young/ha each for *T. longipes* and *T. gracilis.*

The only other field data on the breeding of individual treeshrews is Langham's work on *T. glis.* Most of the fifteen to thirty adult females he was able to capture monthly bred only once in the year, but five older females had two or three litters (Langham 1982). Reproductive output per female may thus have been somewhat lower at his study sites than at Danum Valley in 1990–91.

Life History

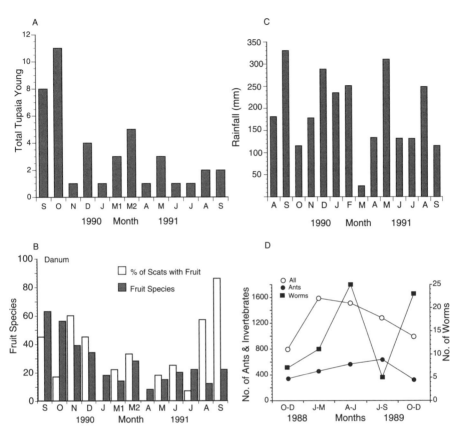

Fig. 10.4. A, the month of first capture of all young of *Tupaia* species at Danum Valley. B, the number of fruit species collected monthly in the transect of study area trails (shaded bars) and percent of treeshrew scats including fruit (open bars). C, monthly rainfall. D, numbers of litter invertebrates collected in standard litter samples during three-month periods in 1988–89. (Compiled from Burghouts et al. 1992 and. T. Burghouts pers. com.)

ENVIRONMENTAL CORRELATES OF REPRODUCTION

Why do treeshrews breed in some months and not others? As in all equatorial rainforest climates, in Sabah neither daylength nor temperature vary sufficiently to influence mammalian ecology, and rainfall is the chief seasonal variable. Rainfall patterns orchestrate the life histories of terrestrial organisms, directly through effects of precipitation and indirectly through proximate control of plant phenological cycles (e.g., Foster 1982).

The patterns of treeshrew reproduction, monthly fruitfall (see also chap. 5), and rainfall (fig. 10.4C) show strong similarities. However, the closest tracking is between emergence of young and fruitfall one to two months earlier. The study at Danum began during one of the great fruit-masting episodes for which the forests of Southeast Asia are famous (Janzen 1974; see also chap. 5). As this was the first major masting year recorded in the six years since the founding of the Field Centre, it was a rare event. Our fruit sampling started in September, when masting was already well in progress, and we are unsure whether the September 1990 peak in fruit species was the maximum or whether it had peaked earlier (fig. 10.4). Maximal treeshrew reproduction for the entire study coincided with the maximal fruit-masting species peak from September to November 1990. In September 1991 the peaks in both fruit and reproduction were much smaller. Appearance of young also followed, two months later, the March fruit peak, which may be the largest in a non-masting year. Peak numbers of captured young seemed to be timed such that females would have become pregnant during fruit maxima, but some young consequently became independent (trappable) during the fruit minima in March and May (fig. 10.4).

Seasonal abundance of soil surface and leaf litter invertebrates next to the treeshrew study plot at Danum from October 1988 to December 1989 show invertebrates to be rarest in October–December (Burghouts et al. 1992) and most numerous from January to June (fig. 10.4). The samples with highest numbers of total invertebrates generally coincide with the March breeding peak of the litter-foraging treeshrews. Numbers of surface earthworms (fig. 10.4) seemed to be directly related to the amount of rainfall, as one would expect (highest in June, July, October, and November in 1989).

GROWTH RATES OF YOUNG

The known histories of all individuals of *Tupaia* captured at Danum Valley can be plotted to show the phenology of body mass (figs. 10.5–10.7). This gives (1) a measure of the growth rates of young in the field; (2) adult body weight changes that could be linked to breeding seasonality or other factors; and (3) the turnover of individuals on the study area, for visual review. The weights of pregnant females are included.

Several juveniles of *T. longipes* and *T. tana* were recaptured as they grew, and five of these remained to adulthood on the study area. In both of these species young born in September or October 1990 showed a very fast weight gain of about 50 to 60 g during about the first month

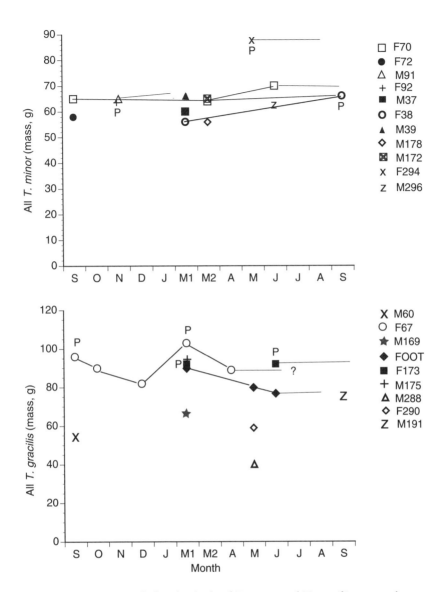

Fig. 10.5. Histories of all individuals of *T. minor* and *T. gracilis* captured on the Danum Valley study area, by captures during monthly trapping sessions. P = pregnant; adults = open symbols; young = closed symbols. Thin trailing lines are periods when the individual was known alive, from radio-tracking or sightings, beyond the last recapture date.

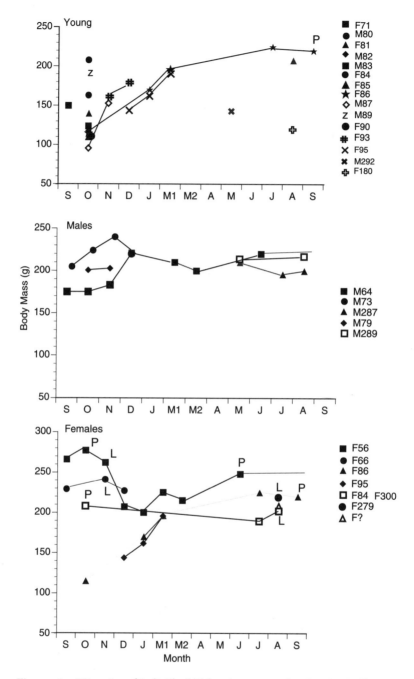

Fig. 10.6. Histories of individual *T. longipes* trapped at Danum Valley. P = pregnant; L = lactating. Adults = open symbols; young = closed symbols.

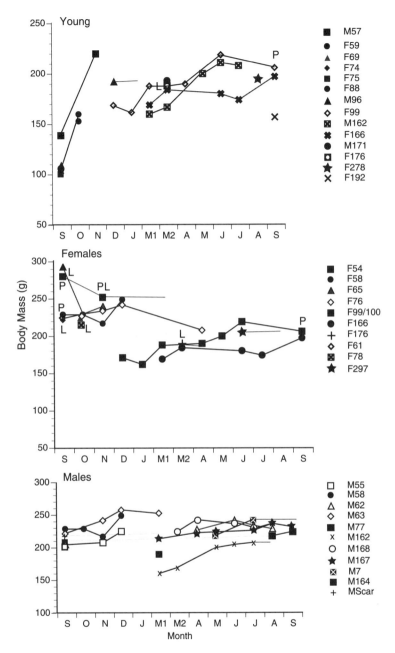

Fig. 10.7. Histories of individual *T. tana* on the study site at Danum Valley. Young <200 g at first capture (top), resident females, including young that became established on territories and that are also shown in the top figure (middle), and adult males (bottom). P = pregnant; L = lactating.

following emergence. Thereafter, weight gain was extremely slow, usually from 10 to 20 g per month, and adult weight was only reached after at least another four to six months or longer. For example, *T. tana* F100 gained 40 g between January and May (162–200 g); F166 gained 30 g between early March and September (169–197 g); and M162 gained 40 g from early March to May (160–200 g) (fig. 10.7). Some *T. longipes* youngsters seemed to gain weight slightly more rapidly than *T. tana*. In the three cases in which weights were obtained in two successive months, each gained about 20 g/mo when they were probably two to four months old. There were too few captures of *T. minor* to show any trends other than a quite uniform body weight, and we collected no data on the growth of young for any of the three tiny species.

AGE AT FIRST REPRODUCTION

Two females each of large and plain treeshrews that were born on the study area at Danum Valley became resident, territorial adults there. Three of these, one *T. tana* and two *T. longipes*, were born in September–October 1990 (F99/100, F81, F86). Two showed their first reproduction in September 1991 and one in August. The fourth female (F166) was probably born in December or January 1991. When last captured in September 1991 she was still nonreproductive. At Poring a subadult female *T. tana* of 144 g (F122), was established by 2 May, following the April death of the resident, but she was still not reproductive in mid-August. These few cases indicate that *T. tana* and *T. longipes* first reproduced when about a year old, even when they had acquired a territory several months previously. The single exception was a young adult female *T. tana* (F176; fig. 10.8) of 188 g, with unworn teeth, which was captured lactating on 30 March. She may have been born the previous August–September. Her radio signal vanished within two weeks, and she was not recaptured, so her fate was unknown.

The captive-bred treeshrews studied by Martin (1968) mushroomed to adult weight (200 g) by two and a half to three months old, only two months after weaning. Sexual maturity arrived at three months and birth of the first litter at four and a half months. There is even a report of a captive female becoming pregnant at two and a half months (Sprankel 1961). Langham's (1982) excellent field data on the weight gain of wild *T. glis* in West Malaysia, with measurements of from thirty to sixty youngsters trapped per month, has a virtually identical pattern to that I recorded at Danum Valley. Adult weight in West Malaysia (mean = 149 g) was reached only after more than eight months. Juveniles from two to five months old only gained about 10 g per month, and only about 5 g per month thereafter (at eight months subadults weighed about 135 to 140 g). In the same study the youngest recorded breeding age for a female was seven months, but, Langham reported, "there is some indication that animals in their first year frequently do not breed either because they find it difficult securing a mate or holding a territory" (p. 332).

SEASONAL PATTERNS OF ADULT WEIGHT

The females of *T. gracilis, T. longipes,* and *T. tana* that were followed for several months all showed maximum (nonpregnant) weights somewhere between September and November and falling weights toward December and January (figs. 10.5–10.7). Some also dipped in weight between April and June. Two subadult *T. tana* lost weight in December and January, and one lost weight in July.

Only in *T. tana* and *T. longipes* were males followed for extended periods (figs. 10.6, 10.7). Like the females, males showed maximum weights in the first three months of the study, but they differed from females in that they peaked later, in either December (N = 3) or November (N = 1). The lighter individuals of each species, M64 and M55, were both young and were apparently still growing to adult weight, but M73 was an old adult and M63 was mature, so their weight increases were due to improved body condition, not net growth. Males also showed a slight tendency to lose weight sometime between March and July.

If the body mass phenologies are compared with fruitfall and rainfall phenology (fig. 10.4), the patterns seem to show the closest similarity to rainfall, rather than to crude fruitfall, but treeshrews in general had the highest body masses during the peak fruiting period of September–December 1991. Too few individuals are represented to give more than a general idea of trends.

PERSISTENCE (SURVIVORSHIP)
OF INDIVIDUALS ON THE STUDY AREA

Treeshrews are for the most part highly trappable, and all individuals of the four terrestrial species studied were initially easily caught; but in any month some animals escaped capture and were not monitored (see chapter appendix). In general, I only assumed that an adult, territorial resident treeshrew was gone from the area when another adult was captured on its former territory. In most months the differential between captures and individuals known alive was small for the three *Tupaia* species for which there is some population data (fig. 10.8).

For all three species, the population of resident adults was completely stable from September to December 1990, but there was a changeover of residents between December and March and into June (figs. 10.6–10.8). In January–February I was absent and could not record the fates of the animals.

TUPAIA GRACILIS AND *T. MINOR*   There was only one adult resident slender treeshrew on the study area at Danum at the beginning of the study and two at the end of it. No individual remained on the area throughout the study, but F67, who was displaced from the study area in March, after ten months, by a new adult female, perhaps was still alive. The population thus increased while two females shared the area formerly used by one (fig. 10.8). The adult male (Mfoot) took up residence in March and disappeared in August, after six months, and a new, barely adult male was captured in September (M191). None of the four subadults was captured a second month.

Although too few individuals were followed for much inference, at least one *T. minor* (F70) remained a resident adult on her territory for the eleven months of the study at Danum. This shows that the smallest treeshrews can be long-lived.

TUPAIA LONGIPES   At Danum Valley persistence of resident adult *T. longipes* was high: three (perhaps four) of the five adults trapped in September 1990 were still present at the end of the study in August–September 1991 (60–80% survival). One old male (M73) with much tooth wear in September 1990 disappeared after December and by June was replaced by a young male with little tooth wear (M287) (fig. 10.6). One of the resident females lost half her territory in March–April to a young female born on the study area in September (see chap. 9). Young of the year thus appeared to squeeze or replace two of the older residents. This division/compression of female territories increased the total population. Two territorial females and one male (F56, F84, and M64) survived the entire study, and the other female (F66) was not replaced and may have been still alive but trap-shy (as was M64 at the end). At Poring the two resident females survived the entire study, but the male was not recaptured at the end and his status was unknown. In this species alone, I caught ancient individuals with teeth worn nearly to the gums (one of each sex), suggesting frequent high survivorship.

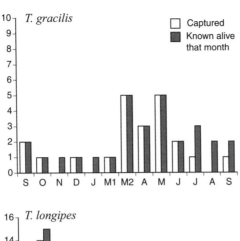
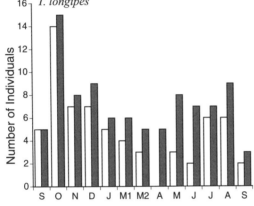
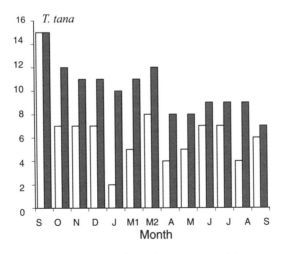

Fig. 10.8. Number of treeshrews captured (open bars) and number known alive (shaded bars) each month at Danum Valley.

TUPAIA TANA Of the nine resident large treeshrews captured in September 1990, only two males (M62, M77) are known to have survived the entire year; one of these (M77) had apparently been pushed out of part of his original territory and was no longer within the trappable sector. One female (F54) was not replaced by another on her territory, so she may have been alive but trap-shy (she was injured in a trap). Survivorship for the year was thus 22 to 33 percent. There was a large turnover of *T. tana* in December–March (fig. 10.7). Most animals simply disappeared and were replaced by others, but one male (M55) was evidently killed by a predator. Two females and two males were replaced by animals born on the study area, all from the August–September generation (fig. 10.7), and two other new males and a female (M167, M168, and F297) were also of this age, as judged from tooth wear.

Langham (1982) collected excellent demographic data during his three-year study of *T. glis*. His study population had a one-year survivorship of 33 percent for males and 40 percent for females and a two-year survivorship of about 10 percent and 12 percent respectively. I followed too few animals for much inference, but my study populations of *T. longipes* seemed to have higher adult persistence than Langham's, and *T. tana* persistence was similar or slightly lower than his, based on the standing estimated age distribution from tooth wear at the beginning of the study (see chapter appendix).

Langham found that most adult *T. glis* disappeared in three periods: in December–February, June–July, and October–November. This is echoed by the strong seasonality of territorial turnover among animals at Danum. Individuals of three species (*T. tana, T. gracilis,* and *T. longipes*) were displaced by other individuals from all or parts of their home ranges. Treeshrews that disappeared from territories were therefore sometimes still alive. I conjecture that the cause of the seasonality in territorial replacement at Danum was the simultaneous growth to adulthood of a super cohort of young born in August–September. Because of the fruit masting, these may have been super-young: unusually fast growing and large, which succeeded in challenging and ousting older, established residents.

In summary, the three terrestrial species all shared some similar life history patterns: (1) breeding in two main periods that followed peak fruitfalls, August–November and March–June; (2) a shift in territory ownership (population turnover) in January–May, resulting in a higher density of resident individuals and slightly smaller territories; (3) slow final maturation of young, so that young of the year that moved into territories during the January–May period mostly began to reproduce in the August breeding season.

DISCUSSION OF LIFE HISTORY

All fieldwork on treeshrews thus shows that the explosive reproductive potential from continual breeding, rapid growth, and early sexual maturity seen in laboratory treeshrews is not realized in the wild, where discontinuous breeding, slow growth, and late maturation are more typical. Because there is no evidence for genetic control of seasonal breeding, I agree with Langham's hypothesis (1982) that in nature treeshrew growth and reproduction are nutrient limited. Langham speculated, "The delay in attaining sexual maturity was probably a result of a decrease in food supply. As most *T. glis* were born in the period March to August, they would be growing to adult weight during the period of the north-east monsoon, associated with heavy rain and lowered food supply" (p. 331). And further: "Breeding was timed so that births tended to coincide with the onset of the south-west monsoon with its moderate rainfall, leading to an increase in vegetative growth and a presumed increase in insect abundance" (p. 341). In Sabah young likewise sometimes emerged when food supplies were at minima.

Langham believed that arthropod abundance was the food variable limiting treeshrew breeding. Soil arthropod abundance in his region was highest in April to August and lowest in November (Langham 1982). This does not correlate well with peak breeding of *T. glis* in February to June, but the sector of arthropods sampled was not the same as those eaten by *Tupaia*. My results point to fruitfall as the proximal factor controlling reproduction, and especially reproductive success, but there are insufficient data on arthropod abundance in Sabah to evaluate its possible role, which is likely an additive one including both fruit and invertebrates.

Several features of treeshrew behavior in Sabah support the conclusion that nutrient levels limit reproduction. The long distances of daily movements, dawn-to-dusk activity, and large home ranges of *Tupaia* species directly imply that these treeshrews need the entire available day and a large piece of the landscape to meet their daily food needs. The concentration of activity at fruit trees and the shortened daily ranges of treeshrews using them (table 8.4; chap. 5) indicates that the animals use fruit when they can, even to the point of overstepping territorial boundaries (Kawamichi and Kawamichi 1979; see also chap. 9). Arthropods are hunted one by one, by searching, searching, and more searching, but abundant fruits such as figs can be eaten as quickly as they can be passed through the gut, without travel expenses. Treeshrews may well be able to maintain themselves on an invertebrate diet, but I believe that some *Tupaia* species need the extra calories provided by fruit to amass energetic reserves for reproduction. The best evidence for this is the response of treeshrews to the fruit-masting phenomenon of September 1990 (fig. 10.4). This coincided with a dramatic breeding peak in *T. tana* and *T. longipes*, when females apparently bred two to three times and two generations of young were simultaneously present. The following September (1991), when one-third as many species fruited, only one-fourth as many young emerged. As shown in Appendix VI, a rodent population also rocketed after the masting. A study of the phenology of fruits eaten by primates in West Malaysia (Raemaekers, Aldrich-Blake, and Payne 1980) shows the highest number of species fruiting in a broad peak from January to June. This exactly corresponds to the *T. glis* breeding season observed by Langham (1982), but a lesser fruit peak in September was not mirrored by treeshrew reproduction. The phenology data are for a different year and locality, so only broad, general trends may be meaningful.

Field data may not discriminate the relative importance of individual food types for reproduction. Seasonal abundances and/or availability of fruitfall, arthropods, and earthworms could be synchronized, as all are keyed to rainfall. If breeding is facultatively triggered by a nutrient and/or energy threshold, then any combination of food items could suffice, such as a mix of high invertebrate prey levels and some fruit, or vice versa. The one-month phase difference in breeding of *T. tana* and *T. longipes* in September–October 1991 suggests either that a nutrient advantage for *T. tana* occurred earlier or that *T. longipes* had a larger energy deficit to overcome before breeding. Because large treeshrews have smaller daily movements and home ranges than plain treeshrews, they have energetically easier lives, which may translate into a lower nutrient threshold to overcome before breeding. *Tupaia tana* females were active for an average of 1.25 h longer per day than males (see chap. 7). This could reflect a greater effort by females to acquire resources for reproduction yet energy acquisition easy enough for males to be able to retire early. Such a differential was not evident in other *Tupaia* species, where both sexes were active for nearly 12 hours a day, suggesting a more difficult fundamental regime (see table 7.1).

Year-round physiological readiness would permit treeshrews to breed rapidly whenever food becomes abundant and to remain close to carry-

Life History

ing capacity at all times. The tenfold higher population densities and comparably shrunken territory sizes of *T. glis* in West Malaysia and Singapore (Kawamichi and Kawamichi 1979; Langham 1982), compared to those of *T. longipes* in Sabah, provide insight into the interplay of ecology and social organization. In the presumed presence of higher resource concentrations in West Malaysia, *T. glis* shrinks its territory size and thereby hugely augments its population density, to the point where its nutrient-determined reproductive parameters of litter number per year and growth rate of young are close to the same, or even below, those of species in Sabah that have to forage huge daily distances on giant territories. The balance between social forces involved in territorial defense (intraspecific competition) and resource supply on the territory evidently does not allow the existence of fat treeshrews with rich territories that allow them to breed continuously. However, the serendipity of a massive fruiting phenomenon allows them to temporarily greatly augment reproductive output, such that while the fruit is available, breeding is continuous.

THE ABSENTEE SYSTEM

The absentee maternal system requires for its success a whole suite of behavioral, physical, and physiological adaptations that are highly specific to it. These include (1) highly specialized visiting behavior by the mother; (2) storage and instant release of forty-eight hours' worth of milk by her mammary glands; (3) intense, virtually instantaneous suckling behavior by the nursling; (4) storage of the milk in a giant stomach; (5) thermoregulation by newborns; (6) the ability of nestlings to maintain rapid growth and resist dehydration; and (7) energy-minimizing juvenile behavior, such as absence of exploratory behavior. The absentee nursing schedule must tightly constrain litter size. To succeed in weaning more than two young, the mother would probably have to increase the rate of feeding to at least daily. It was certain that this system had to occur in the wild, because captive treeshrews rear young brilliantly with it.

The absentee maternal care that we documented in *T. tana* F109 corroborated in detail the behavior seen by Martin (1968) in captivity. This includes an "extra" nursing visit on the day before emergence, which itself was on a normal feeding date. I surmise that this one extra feeding gave the young enough surplus energy to awake from their usual torpor and venture out into the world. The feeding supplement itself may have stimulated a change of activity pattern.

The most unexpected and novel of our findings on *T. tana* maternal behavior in the wild was not that she spent only two minutes every other day with her nestling young but that when these left the nest she abruptly gave them enormous amounts of attention (Emmons and Biun 1991). *Tupaia* had been viewed as having the least possible maternal care for a mammal, with barely the rudiments of behaviors that are normal in other nidicolous taxa (Martin 1968). Our observations of a wild treeshrew imply that this is not so and that treeshrew mothers can devote much time to postemergence care of their youngsters. The regularity of the early-morning visits of the *T. tana* to her newly emerged young suggests that she may have continued to nurse them, and the many hours she spent with them suggest that she may have helped them learn to forage (Emmons and Biun 1991). The mother *T. tana* both groomed her young and provisioned them with solid food when they first emerged from the nest, so these behaviors do exist in a tupaiid. The long hours and long distances needed for foraging by adult treeshrews indicate difficult lives, and newly emerged young may need help to survive the weaning phase. *Tupaia tana* seems to have a reversed behavioral investment sequence compared to most nidicolous mammals, which invest heavily in physical care of young nestlings, then abandon them as soon as they emerge.

Tiny pentail young shared the den of the mother at about the time of their emergence, and evidently continued to do so for months, probably until after the next litter was born, as we saw adults, subadults, and juveniles emerge from the den together. This, coupled with other evidence that pentail groups share a nest (see chap. 9), makes it almost certain that pentail treeshrews do not have an absentee maternal care system. If this most primitive treeshrew has a "normal" nidicolous system wherein mother and young nest together from birth, then it is probable that the absentee system of *Tupaia* is a derived condition in the subfamily Tupaiinae. If ancestral tupaiines switched to an absentee system, what might have provoked the change?

When he described the maternal care system of *Tupaia*, Martin (1968) speculated at length about its evolution and possible function. He thought that only two hypotheses were likely: (1) that it functions to protect young from predation and (2) that it has no specific contemporary function but is a relict, primitive system that is carried by treeshrews as evolutionary baggage. The two hypotheses are not mutually exclusive, but Martin favored the second hypothesis and rejected the first, because he believed that treeshrew nursery nests were more likely to attract pred-

Life History

TABLE 10.5. Schema of possible functions of absentee maternal care and the probable "reason" for the system in other mammals.

	Selective Force	
Force acts on	Predation	Energetics
Young	Seals Tupaiids Lagomorphs Agoutis Artiodactyla Elephant shrews?	Seals (energy storage) Tupaiids
Mother		Seals Tupaiids?

ators than are normal nests of nidicolous mammals. In his captive colonies, the nursery nests were smelly and soiled with urine.

An approach to this question is to explore whether tupaiines share common features with other species that have absentee or similar maternal care systems (table 10.5). Mammals with absentee-like maternal care systems in which the young rest apart from the mother, who visits them for nursing, include taxa as diverse as elephant shrews, hares, pikas, maras, agoutis, most deer, many antelopes, and some seals. Apart from the seals, most of these are both cursorial and have precocious young (Emmons and Biun 1991), and this includes elephant shrews and many lagomorphs. The connection between these features is obvious: cursorial mammals have long legs and modified feet and rely on rapid getaway for survival from predators. Their limbs are poorly suited for escaping within narrow burrows or nests, or climbing anything, and they generally rest in sites from which cursorial escape is possible. Their young must either be precocious enough to outrun predators shortly after birth, as in the case of horses, caribou, or wildebeest, or they must be hidden distant from the mother, to which predators are attracted, as in most small deer and antelopes. Small cursorial species whose legs are unsuited to burrowing must hide their young in the open, but saltatorial species that jump from the hind legs, so that the forelegs are less modified for running and can be used for digging, such as maras, agoutis, and some rabbits, can dig a burrow for their young but do not use it themselves, as they are probably safer outside it. Elephant shrews use their forelegs for digging in the litter, but only one species burrows. The others use shelters under the litter (Rathbun 1979), a possibility only for mammals small enough to hide under fallen debris. In all of these cursorial species, the absentee-like maternal behavior is clearly a means to hide the young from predators while the conspicuous mother (which can run from them) forages at a distance. All these other species feed their young more often then do *Tupaia*.

A type of absentee maternal behavior is exclusive to the seals and sea lions, especially the eared seals (Otariidae; the following from a review in Oftedal, Boness, and Tedman 1987). These give birth on land in protected sites (usually islands), where the mother leaves the young while she goes on extended foraging trips that can last for more than ten days but usually two to three. When she returns she may stay with the young and nurse it for a day or two before leaving again. Lactation is prolonged in these species (6–12 months). In phocid seals, which place their young on pack ice, the mother stays with the young and does little or no foraging until the young is weaned after a short lactation period (4–30 days). The absentee pinniped lactation systems are driven proximately by energetics because young on land or ice are often so far from the foraging grounds that several days are needed for the mother to forage. Ultimately, they may be driven by predation avoidance, which favors the use of safe, remote nurseries for the young.

Most terrestrial species with absentee-like maternal care are diurnal, or active crepuscularly or both day and night. This may imply that visual detection by predators of mothers with their young is a more important problem for diurnal than for nocturnal species. Diurnality is quite rare among small mammals, and it may be no coincidence that elephant shrews, tupaiines, and pikas share both diurnal activity and absentee maternal care. By day predators may be more visual and able to spot maternal visits to young from a greater distance. This could select for reduced nursing visits.

Tupaia species differ from all other absentee mothers in that the adults use nests, at least one species is arboreal, and the young are altricial. Because hidden-young systems have evolved independently many times as a protection against predators, it is also likely to have this function in treeshrews, despite the different ecomorphology. A reason that predators would be less likely to find young in absentee nursery nests is that adult treeshrews have a very strong musky odor (especially their urine). When we removed them for examination, I sniffed both the *T. tana* nursery nest and the young. Neither had any perceptible odor, and they were clean and dry. If the mother shared the nest, it would have a stronger scent when she was present, and probably also when she was not. Parasite loads should also be lower on young isolated from adults in separated nests.

The absentee system of *Tupaia* could also confer an energetic advantage compared to "normal" nidicolous systems. Energy would be saved by the mother avoiding the commute several times a day from her foraging place to the nestlings, and more would be saved by the extreme immobility of the young nestlings, which because of their forty-eight-hour fasts must hoard the energy from each feeding. The whole system is perhaps more energy efficient, which could be critical in these energy-limited species.

Some hypotheses about the systematic place of treeshrews put them into the same order with elephant shrews, or rabbits and pikas (reviewed in Luckett 1980). The morphological basis for these ideas is weak (Luckett 1980), but it is curious that these divergent (from other mammalian families) and presumably old orders share with treeshrews an absentee maternal care system (Broekhuizen and Maaskamp 1976; Emmons and Biun 1991; Rathbun 1979; Smith and Ivins 1983). Is this evidence for an extremely early origin of this system in some remote common ancestor of these families? A brief review of what is known of the most primitive living mammals does not suggest that these have absentee or any less intense or more rudimentary maternal care than do members of more recent, "advanced" families (Emmons and Biun 1991): even platypus mothers stay with their young in the same nest for many weeks (Grant 1983).

APPENDIX: METHODS

Maternal Care. I knew that finding nests with young was going to be most difficult, because locating a single spot visited by an animal for less than five minutes during two twelve-hour days of activity, by means of radio-location alone, is equivalent to finding the proverbial needle in a haystack. At the outset it was not even known where most of the species nested, or what nests of any species looked like. My method for looking for nursery nests was straightforward: I would follow a lactating, radio-collared female for four days, then map her paths and look for places she went in the early morning on alternate days, or each day. If I could pinpoint a likely spot, I would track her to see if she went there on schedule on an additional day. Before daylight on the next appropriate day, I would try to hide in a place with a view of the surroundings and monitor the signal through earphones as the female approached. Despite much effort, this never worked. It was a time-intensive effort, because it took about ten days for each attempt. The second year of the study, I tried another method: because females fed their young soon after daylight, I attached a spool-and-line device (Ryan, Creighton, and Emmons 1993) to a lactating female and released her at nightfall. She removed the spool overnight, but she was evidently disturbed by the procedure and greatly delayed leaving the nest in the morning, so I did not repeat the experiment (treeshrews travel so far in a day that they would use all of the thread that they can carry in a few hours, so to find a nest with young, one would have to attach the spool the night before a nursing). A suggestion for future efforts would be to try marking the female with fluorescent powder.

To observe the treeshrew nursery nest that we found at Poring, we built an elevated platform in a tree, hidden by a blind of palm leaves on all sides (Emmons and Biun 1991). One of us would take up position in the blind well before dawn and monitor the radio signal (with earphones) from the mother, whose own nest was 70 m away. When she approached, we turned off the receiver and waited motionless until after she had visited and moved well away. We videotaped her arrival on one morning. During an "off" day, when the young were not nursed, we climbed up and briefly removed them for measurement and to mark them with ear tags and describe the nest. At that time we estimate that they were eighteen days old. One was male and the other female, and their measurements were as follows: total length = 206 mm; tail = 82 mm; hind foot = 37 mm; ear = 10 mm; mass = 54, 55 g.

Life History. Life history data were derived largely from monthly capture records. At first capture each month, all treeshrews were weighed, and females were palpated for the presence and number of embryos and examined for evidence of lactation. To avoid physical handling, I did not measure them except at first capture or if they were anesthetized for fitting with a radio-collar. Ideally, monthly trapping would be scheduled so as to capture almost all of the resident animals, but in reality it was a compromise where several factors needed to be juggled. *Tupaia* species often became trap-happy. Some animals were caught twice daily and consequently spent almost the whole day in traps. To decrease the likelihood of injuries to these, and minimize disruption of their lives, monthly trapping was limited to two days. This was too brief to capture every treeshrew. In contrast, a few individuals, especially *T. longipes,* eventually became trap-shy after multiple bouts of handling, so that when their radios failed, I usually lost track of them, although some were seen from time to time so I knew they were still on their territories. Monthly records were therefore incomplete.

Capture Effects. Of great concern when doing research in which animals are submitted to trapping, handling, anesthesia, and carrying radio-collars is whether the procedures of the study cause stress or mortality that change the life histories of the animals. One way to evaluate whether the population is in a "normal" state after a year of study is to compare its age structure at the beginning of the study, when animals were first trapped, to that at its end. This can be done roughly, for *T. longipes* and *T. tana,* from notes on tooth wear. In September 1990, of six adult *T. longipes* caught, four were old, with very worn teeth, and two were just entering adulthood. At the end of the study at least three (perhaps four) plain treeshrews on the area were very old, and two were young of the year. *Tupaia tana* residents on the study area in September 1990 included five older animals with worn teeth, perhaps eighteen months to two years old, but none with very worn teeth, and five younger ones, probably young of the year. At the end of the study there were only two (possibly three) old animals and six young ones. The population structure of plain treeshrews was thus unchanged during the study year, but large treeshrews appeared to have a somewhat higher turnover of individuals. However, because of the rare fruit-masting episode, the study period was not a "normal" year, and unusually high numbers of dispersing young in March may have accelerated territorial turnover.

CHAPTER 11

Predation, Predators, and Alarm Behaviors

13 Apr. 1989: [While *T. minor* F112 is eating fruits in a treelet canopy] a large butterfly or bird flies over, and she takes a leap undercover of some leaves.
    18 Oct. 1990: 0639 h. A *Sundasciurus hippurus* [horse-tailed squirrel] alarms above and a *T. minor* drops down lianas with a chatter.
    9 Nov. 1990: 1201 h. I scared a *T. minor* that was in a thicket at 2–3 m. It chattered and dropped into the undergrowth.

The chief antipredator strategy of treeshrews is to avoid being noticed. They do this with silent, inconspicuous movements and by keeping undercover, helped by their cryptic coloration. Little escapes their acute senses and extreme alertness. When a treeshrew spots a suspicious but not immediately dangerous situation, such as a motionless scientist, it is likely to give alarm calls from a discreet distance and safe site, such as a log, brush pile, or vine tangle. If the same threat is moving along, the treeshrew is unlikely to call but freezes watchfully or vanishes silently. Pentail treeshrews sometimes dashed around tree trunks to hide from me on the opposite side. I never saw lesser treeshrews do this, but they were rarely on trunks. Pentails often ran higher up into trees when they saw me. Lesser treeshrews seemed to have little anxiety about persons below them but reacted strongly to threats from above, which caused them to drop downward even if they were close to the ground and the threat (me) was terrestrial. I saw lesser treeshrews carry both fruits and insects to eat in bowers screened by leaves from above or crouch in such shelters to rest.

All treeshrews moved so quietly that I rarely heard them. Even during dry spells when the leaf litter crunched like cornflakes underfoot, terrestrial treeshrews traveled noiselessly. They achieved this by taking itineraries along logs, roots, vines, saplings, and fallen branches, avoiding the dry litter, as well as by keeping to the danker, denser vegetation that was sheltered from desiccation. Thus the scansorial morphology of ground-foraging species has an unexpected safety function in facilitation of quiet travel entirely at ground level along "arboreal" substrates deployed on the ground.

Lesser treeshrews also moved silently while actively foraging in foliage. This was accomplished by a careful, somewhat sneaky and weasel-like motion along interior supports, with little jumping. However, their movements caused quivering of outer foliage, so that I could see where they were working hidden from view inside vine tangles. Whereas squirrels jump noisily between terminal branch tips or bound across dry leaves, treeshrews move with discretion and pass unnoticed.

ALARM CALLS

Each treeshrew species can be identified by its distinctive alarm calls. Each has at least two types of calls that seem to signal different levels of arousal. Lesser treeshrews make a soft birdlike peeping when slightly aroused and a loud, squirrel-like staccato chatter when more strongly alarmed. A slender treeshrew was noted peeping only once (a known, marked animal) but commonly called with repeated long soft whines. *Tupaia longipes* was also only once heard peeping and more usually gave long rasping whines, or sometimes hoarse squawks. Montane treeshrews make a squawking sound reminiscent of this latter call of *T. longipes*. Large treeshrews chatter when surprised at close range, but if they spotted me from a distance they called with about three warbling birdlike whistles followed or not by chatters. Rarely, this species also squawks. Pentails make soft to loud birdlike chirps and chirrups, raspy to musical. The most similar calls were the long whining notes of *T. longipes* and *T. gracilis* and the squawks of *T. montana* and *T. longipes*. These three species also shared a loud, unnerving distress scream, produced when constrained in the hand, the like of which was never elicited from large or lesser treeshrews.

Alarm calling was accompanied by an upward flick of the tail to vertical, but alarming terrestrial treeshrews were not very conspicuous (unlike squirrels), and the calls were of low amplitude and hard for me to locate. The long, low frequency whines of lesser and plain treeshrews are of a structure thought to be difficult for predators to pinpoint (Morton 1975). Alarm calls of lesser treeshrews, from the safety of trees, were louder and of a structure that was easy to locate (a loud chatter), and the animals were often in plain sight. Like squirrels, suspicious or uneasy treeshrews would flick their tails without calling. Alarm calling by treeshrews was not contagious, unlike the mobbing calls of squirrels, which can attract other calling squirrels of the same and different species, as well as birds and monkeys (Emmons 1978). Of seventy-six calling episodes noted in the field, only three included more than one tupaiid. In one case each, two individuals of lesser, plain, and pentail treeshrews called near each other.

Terrestrial treeshrews were not in general very vocal and called rarely, with the exception of *T. montana,* which was the only species often detected through its calls. In contrast, the two arboreal species called often: lesser treeshrews often peeped softly as they foraged; perhaps peeps were contact calls between pairs. Pentails usually called when they saw me near their nest tree. The most intense pentail calling bouts were those of the adult female when small juveniles had just emerged. Several pentails were always likely to be outside near the nest tree simultaneously, because they entered and exited the den at the same time, so calling at the den tree was likely to be heard by other group members (see figs. 7.1, 7.2).

There is a curious similarity between the whining alarm calls of lesser and plain treeshrews and that of the terrestrial pittas (*Pitta baudi, P. venusta,* small forest birds) that intimately share their microhabitat. Before I knew them well, I sometimes mistook a faintly heard call of one taxon for the other. Once, seeming to answer, a pitta instantly called after an alarm of *T. longipes,* perhaps also mistaking the caller (birds are easily tricked to respond to imitated calls). One can speculate that this resemblance between bird and treeshrew calls is a useful convergence between animals that share the same microhabitat and predators; but birds and mammals worldwide readily learn to respond to each other's alarm calls without convergence in call type (when a squirrel or jay alarms, every monkey, bird, or deer is instantly alert).

PREDATION

As often in field studies of small mammals, we did not witness any acts of predation on our study animals. There was strong evidence for only one case: the radio-collar of a male *T. tana* (M55) at Danum Valley was recovered from a brush pile, with the brass band of the collar crushed flat and deeply scored by tooth marks, as if compressed in a killing bite to the neck by a mammalian carnivore. There was no trace of the body, and possibly the animal died of other causes and the collar was then damaged by a scavenger. We found undamaged, empty, radio-collars at other times; usually the treeshrews had slipped them off (the animal was recaptured), but the fates of a few were unknown. In an unnatural predation attempt, a trapped juvenile *T. longipes* was attacked by a snake (*Elaphe flavolineata*) that had passed through the mesh (I rescued it in the nick of time).

PREDATORS

Treeshrews have many potential predators. My sightings in each study area (table 11.1) show that yellow-throated martens (*Martes flavigula*) were the most numerous diurnal mammalian carnivores. These large (1–2 kg), muscular martens hunt swiftly, sometimes traveling in pairs or triplets. They sniff intently as they ferret along the ground investigating every treefall, or climb up and down over the boles of large trees. Martens range through the high canopy or the viny subcanopy, leaping with agility from branch to branch. Their behavior makes them especially likely to encounter *Tupaia* nestlings hidden in exposed leaf nests or vulnerable hollows. One morning I was astonished by the sight of a bright orange Malay weasel (*Mustela nudipes*) climbing up to 4 m to forage diligently in a dense vine canopy. It hunted around intently among the vines, descended to the ground, then climbed another vine to repeat the process. This purportedly terrestrial species (Payne, Francis, and Phillipps 1985) may have been searching for birds' nests (22 Aug.), but with this behavior it could discover aboveground nestling treeshrews. The only two Malay weasels that I saw were hunting in daytime (1110 h, 1710 h). Small cats are the mammalian predators most likely to ambush a vigilant *Tupaia* adult. I saw only marbled cats (*Pardofelis marmorata*) active by day and leopard cats (*Prionailurus bengalensis*) by night, but sightings were too few to draw conclusions about felid activity. Clouded leopards (*Neofelis nebulosa*) were reported during the study on roads near Danum Valley, and we found a scat of this species on the Poring study plot (it had eaten a bearded pig). Most of the other mammalian predators would be unlikely to capture active adult treeshrews (table 11.1), but they would eat nestlings if they happened upon them, or inactive adults if they could corner them in a nest.

TABLE 11.1. Number of recorded sightings of Carnivora and other mammalian predators in the forest on or near the study areas (excludes many animals seen on roads or around houses).

Species	Poring	Danum
Echinosorex gymnurus, moonrat (N)		3
Martes flavigula, yellow-throated marten (D)	4	12
Mustela nudipes, Malay weasel (D)	1	1
Mydaus javanensis, Malay badger (N)		1
Prionailurus bengalensis, leopard cat (N)	3	1
Profelis marmorata, marbled cat (D, N)		3
Arctictis binturong, binturong (D)		1
Arctogalidia trivirgata, small-toothed palm civet (N)	3	1
Paradoxurus hermaphroditus, common palm civet (N)	1	17
Hemigalus derbyanus, banded palm civet (N)		6
Prionodon linsang, banded linsang (N)	1	
Viverra tangalunga, Malay civet (N, D)		13
Herpestes brachyurus, short-tailed mongoose (D)		6
Herpestes semitorquatus, collared mongoose (D)		5
Herpestes sp., unidentified (D)		2
Macaca fascicularis, long-tailed macaque (D)		5
Macaca nemestrina, pig-tailed macaque (D)	13	4

NOTE: Little time was spent in the study areas at night, especially at Poring, so nocturnal sightings are few. Period of sightings: (D) = diurnal; (N) = nocturnal.

The carnivore communities were different on the two study areas (table 11.1). Poring lacked Malay civets, banded palm civets, and mongooses. These terrestrial species may have been either eliminated by hunters (poaching was intense) or killed by the domestic dogs that often entered the research area, either directly or indirectly by diseases they carried. However, I saw more small-toothed palm civets and leopard cats at Poring than at Danum, and the study area had a resident troop of pig-tailed macaques (which feed on some small vertebrates). At Danum the study plot had resident long-tailed, but no pig-tailed, macaques, although these occurred elsewhere nearby.

Reptiles, notably snakes and monitor lizards, are potential predators of unwary or nestling treeshrews. Large monitors (*Varanus* cf. *salvator*) feed on both arboreal and terrestrial prey, and I saw them climbing around in the branches of trees, being mobbed by hysterical birds. Thirty-five species of snakes are known from Danum Valley Conservation Area (Anon. 1993), of which at least a third might eat small mammals (J. Cadle pers. com.).

Finally, birds of prey are likely to be the most serious threat to active treeshrews. Of the nineteen diurnal raptors and six owls thus far recorded from the Danum Valley region (Anon. 1993), seven raptors and three owls may take small mammals (Smythies 1981; J. T. Marshall pers. com.). I saw an eagle-owl (*Bubo sumatranus*) perched at 3 to 4 m on a branch above a trail at dawn (0630), scanning the forest floor as if hunting. At that hour treeshrews would have been active for 30 minutes. Thus even owls might take diurnal species. Treeshrews in nature are surrounded, day and night, by a host of vertebrates that would seize an opportunity to make a meal of them. Some invertebrates, such as ants, could also feed on nestling treeshrews.

DISCUSSION OF PREDATION

Lesser treeshrews are nearly indifferent to threats from below, but they react strongly to threats from above, which are most likely to come from birds of prey. Their sudden flight reaction is to drop downward into dense vegetation, even when surprised by a terrestrial threat when they are already very low. The orangutans, red leaf monkeys (*Presbytis rubicunda*), and two species of macaques at Danum all shared this behavior of diving downward when escaping humans. Dropping is a more effective primate strategy for quickly hiding than is escape across the canopy. Treeshrews are likely to drop as an escape from avian predators, but in Borneo large primates have no avian predators, so unless the behavior is a historical relict from a time when large eagles were present (see, e.g., Goodman 1994), the behavior in primates and treeshrews is in response to different dangers. I ascribe this dropping reaction to humans, which is not seen in most South American or African monkeys, at least partly to the lack of a closed canopy or connections between trees that enable horizontal escape in dipterocarp forests (see chap. 3), but 20,000 to 30,000 years of predation by man may also have shaped this behavior.

Some treeshrews nest in poorly protected sites: exposed in treelets; in soft, rotten stumps; or on the ground surface (*T. tana, T. gracilis, T. montana*), while others are well protected in deep burrows or holes in living trees (*T. longipes, T. minor, P. lowii*). Slender treeshrews reacted quickly to disturbance at night by jumping out of their nests, calling in alarm (see chap. 6), evidence that their nests were not safe refuges; but treeshrews in hollow trees, logs, or burrows sat tight regardless of the amount of disruption. The little data collected on longevity (see chap. 10) indicated that *T. longipes,* safe at night in burrows, had a higher adult survival rate on the study area than did *T. tana*. Perhaps predation on inactive treeshrews in their nests is an important mortality factor for species with risky nest sites.

The risk of predation to treeshrews in their nests is impossible to quantify, but young nestlings would obviously be in greater danger than adults, both because they cannot escape and because they are in the nest all of the time. Yellow-throated martens, which hunt by day, would take young, but not adult, *Tupaia* from nests. My observations of martens diligently sniffing their way across all levels of the forest suggest that the forty-eight-hour avoidance of the nursery nest by mother tupaiids, and the cautious, tree-to-tree jumping approach to it that we saw in *T. tana*, could avert discovery of nestlings by these energetic and probably highly efficient carnivores, as well as by other predators. As we reported previously, to us the nursery nest and nestlings of *T. tana* were almost odorless while adult *Tupaia* smelled strongly (Emmons and Biun 1991). The absentee system as we saw it exemplified in *T. tana* should reduce olfactory cues by which predators could discover the young.

PREDATION AND SOCIAL ORGANIZATION: INSIGHTS FROM TREESHREWS

The tendency to call in alarm was greatest in the most social species and those with small territories and dense populations: *P. lowii*, which nest in groups; *T. minor*, in which pairs often forage together; and *T. montana*, which live at highest densities and which during our brief study also included a pair that foraged together all day. In these species calling was most likely to be heard by a mate or offspring. Pair foraging by *T. minor* (see table 9.1), the only diurnal, arboreal species, is likely to be a predator defense behavior, because traveling pairs were hunting invertebrates in such a way that no obvious foraging advantage could accrue from their association. Moving foragers rummaging through foliage must be especially visible and vulnerable to raptors. Traveling pairs of lesser treeshrews emitted frequent peeping vocalizations, perhaps contact calls, and seemed quick to give alarm chatters.

An association between diurnality and cohesive social grouping in small primates has been noted for decades (e.g., Bearder 1987; Charles-Dominique 1975): nocturnal primates are all solitary foragers, whereas diurnal ones all live and forage in social groups. Tupaiids provide a glimmer of insight about this difference. Antipredator behavior may transform a solitary-foraging, faunivore-frugivore at least partially into a socially cohesive, pair-foraging one, with both arboreality and diurnality as contributing or perhaps essential factors. The arboreal nocturnal treeshrew (pentail) foraged strictly alone, despite strong cohesion in nesting groups, while the three diurnal, terrestrial lowland species both foraged and nested alone. The arboreal and diurnal lesser treeshrews maintained solitary nesting habits but often foraged in pairs. This segregates the sociality into a strictly foraging associated activity, with the nesting and absentee system (in captivity) of lesser treeshrews like those of solitary-foraging terrestrial species.

CHAPTER 12

Synthesis

DIET

DIET AND ECOLOGICAL SEPARATION BETWEEN SPECIES

The six species of treeshrews each use the rainforest environment in a unique way (fig. 12.1). Three species each segregate sharply from all the others by major differences in activity phase (*P. lowii*), foraging height (*P. lowii, T. minor*), or habitat specialization (*T. montana*). The three others (*T. gracilis, T. longipes, T. tana*), which coinhabit the lowland forest floor, differ in body size, foraging substrate, and invertebrate prey, as well as in nest sites.

Species of the two pairs that are most similar in body size, *T. minor* and *T. gracilis* and *T. montana* and *T. longipes,* are completely separated by height or habitat, such that they scarcely interact. An initially surprising finding was that mountain treeshrews are most similar in diet, not to plain treeshrews, which they closely resemble in size, nest sites (below ground), and alarm calls, but to large treeshrews, which they do not superficially resemble at all (but to which they are phylogenetically related; Han, Sheldon, and Stuebing n.d.). Only these two species eat earthworms and cryptic arthropod prey dug from beneath the litter (centipedes, millipedes), but they do so at different elevations. Likewise, the pair *T. gracilis* and *T. longipes* are ecologically closer to each other than either is to other species. Both are surface foragers with the same habitat, height, and activity, but they differ sharply in their invertebrate prey choice.

The six species, in the field in their sympatric assemblages, are thus

Synthesis

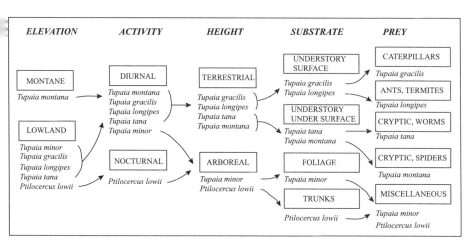

Fig. 12.1. Schematic diagram of the axes of ecological separation between the six treeshrew species studied.

distinct ecological entities, each differentiated from all others by several features. Species pairs of most similar body weight tend to be the most widely separated along microhabitat dimensions. This suggests that current and/or historic competitive interactions have played a role in structuring the treeshrew community. Moreover, it would have been impossible to predict how this community of similar congeners with similar diets (fruit and invertebrates) was ecologically arranged without intensive, species-by-species field study. Many lifestyle differences between the species can most easily be explained or attributed to their differences in foraging sites and prey. Morphological studies likewise could not have predicted many of the ecological differences between species, although they could have detected some, such as the degree of arboreality.

DIET AND DAILY RANGING PATTERNS

FAUNIVORY As discussed in chapter 8, home range size in treeshrews is not predicted by body size: among the three syntopic terrestrial species of *Tupaia, T. gracilis, T. longipes,* and *T. tana,* the exactly inverse relation occurs, with the smallest species using the largest home range and the largest species the smallest. The two morphologically almost "sibling" species, *T. minor* and *T. gracilis,* have, respectively, the smallest and the largest home ranges of the species studied. This is counter to ecological theory and begs an explanation. Because the treeshrews seem to

TABLE 12.1. Summary of ranging patterns of females and foraging characteristics of treeshrew species (from chaps. 4, 5, 7, 8). Hours active = total time out of nest, minus resting time.

Species	Home Range (ha)	Distance/ Day (m)	Hours Active	Foraging Height	Foraging Surface
P. lowii	6.4	1,376	8.7	Arboreal	Trunks
T. minor	1.5	851	10.7	Arboreal	Leaves, vines
T. gracilis	9.1	1,559	11.0	Understory	Surface
T. longipes	6.9	1,800	10.9	Understory	Surface
T. montana	2.5	859	11.0	Ground	Subsurface
T. tana	3.3	1,009	10.5	Ground	Subsurface

feed on the same fruit species, which from their perspectives have similar distribution on the landscape, the substantial differences among treeshrew ranging patterns are likely to result from factors other than the spatial distribution of fruit sources. A comparison of some features of foraging and ranging patterns shows an apparent link between differences in ranging pattern and invertebrate prey foraging modes (table 12.1). I believe that the demands of particular prey capture modes directly result in the ranging pattern differences and their consequences, such as territory size and population density.

The sixfold larger home range of T. gracilis, compared to that of T. minor, implies that its resources are much more thinly scattered in space. An obvious hypothesis to explain this is that whereas T. gracilis is terrestrial, and mostly forages in a two-dimensional plane less than 3 m deep, T. minor uses the whole vegetation column and forages in a 40 to 60 m deep, three-dimensional space that is, moreover, the main production layer of the forest for both fruits and insects. The foliage foraging substrate available to T. minor on its 1.5 ha home range may in fact be as large or larger than that used by a T. gracilis on 10.5 ha. The difference in daily foraging paths of the two species (see fig. 8.4) supports this idea, with lesser treeshrews exploring their areas with much greater relative intensity than slender treeshrews (when depicted as a two-dimensional map). I saw lesser treeshrews spend hours foraging in the vertical dimension of giant, vine-enveloped trees, over horizontal footprints of a few square meters, whereas foraging slender treeshrews kept moving horizontally across the landscape.

The other arboreal species, P. lowii, has a home range much larger than that of T. minor, however, so arboreality per se does not mandate small home range size. The different substrates these two forage on perhaps explain their ranging differences: *P. lowii* characteristically forages on trunks, which do not fill much surface space, while *T. minor* forages on foliage, which does.

Similarly, *T. tana* and evidently *T. montana*, which forage both on the surface and in the subsurface by digging, and use resources from an extra physical layer, have much smaller home ranges than either *T. longipes* or *T. gracilis*, which forage on the surface and have much longer daily movements, larger home ranges, and lower populations than any other treeshrews. These simple (perhaps simplistic) hypotheses will not be easy to test, and other explanations for at least some of the species' home range differences are offered below.

FRUGIVORY   Even if all treeshrews eat nearly the same fruit species, fruit distribution could influence home range size if some species are more dependent on fruit than others. Lambert and Marshall (1991: 793) argued that figs are keystone resources for Malaysian frugivorous birds but that because of the wide dispersion and unpredictability of fruiting, "for frugivores to be dependent on figs in South-east Asia they need to be wide-ranging. Frugivorous birds with small home ranges may have to rely on keystone plants other than *Ficus* during some periods of fruit scarcity." The most wide-ranging treeshrews (*T. gracilis* and *T. longipes*) seemed to concentrate on figs when they could, but so did the least wide-ranging species (*T. minor*; see table 8.4). Fruit eating can reduce daily movements and increase the rate of energy intake (chap. 8), both factors that could shorten daily ranging, but it could ultimately require a larger overall home range to include a quorum of trees. A quantitative difference in frugivory is therefore likely to cause ranging differences among species. The expected pattern with more frugivory would be of larger home ranges relative to daily movements. I have insufficient data to test this, but the small samples collected (see fig. 5.6) point in the right direction, with *T. longipes* and *T. gracilis* showing 40 and 37 percent fruit occurrence in scats but *T. tana* only 28 percent (there are too few samples to consider for the other species). The corresponding travel distance/home range area relationships (fig. 8.3) show that the species with the largest home ranges travel relatively shorter daily distances *per unit area of their home range* than do species with smaller territories. Because each species differs in its body size and insect foraging guild, it is not yet possible to tease out the importance of single factors in controlling home range size, but there are intriguing possibilities to investigate.

COMPETITION   We saw at bait stations that in circumstances of direct contest, a larger species always dominated a smaller one (see chap. 5). This could also bear on the contrasting ranging patterns of the two smallest *Tupaia*. *T. minor,* the only arboreal congener, not only has first pick of canopy fruits while they are on the tree and most fresh and abundant, but no competition for them from other treeshrews (although it does have from birds, monkeys, and Prevost's squirrels), while *T. gracilis,* the smallest terrestrial species, has two larger congeners (as well as other taxa) to compete with for the remainder of the fruit that falls to the ground. For fruit, therefore, slender treeshrews are competitively maximally disadvantaged, and lesser treeshrews are maximally favored. Competition for food resources, with the largest species always favored, could contribute to the inverse pyramids of body and home range sizes that were observed among Sabah treeshrews.

DIET, ENERGETICS, AND SOCIAL ORGANIZATION

Treeshrews in captivity have faster growth rate, earlier maturation, and up to nine times as many litters of young per year as do the same species in the wild (see chap. 10). Wild *Tupaia* therefore reproduce far below their physiological potential. Food shortage is the only likely factor that would limit reproduction in nature (we saw in chapter 7 that time lost from rainfall was unlikely to be significant). Treeshrews of all species were active and presumably foraging for almost the whole day (table 12.1). They thus live close to the limit, in that there is almost no room to expand their foraging time if food supplies become any scarcer. The females apparently reproduce whenever food becomes seasonally abundant, using extra resources for the production of young (chap. 10). The coincidence of breeding with fruitfall peaks, and peaks of fruit in the diet as found in scats, indicates that the extra nutrients for reproduction may often (perhaps always) be supplied by fruit, even though all tupaiids are extensively, perhaps chiefly, faunivorous.

Generalized insectivory is one of the most demanding of (nonbat) mammalian lifestyles, condemning its practitioners to almost constant foraging. Insects are scattered, so that insectivores are likely to do best while feeding alone. Fruit, in contrast, can come in large sources where an animal can gorge itself for several days in a predictable spot, with little physical movement, and fruit sources can often be exploited without sacrifice by many individual frugivores at once. Frugivores are thus often social, living in groups or flocks (monkeys, coatimundis, peccaries), but generalized insectivores are, with rare exceptions (some savanna mongooses that group for defense against predators), solitary or pair foragers (prosimians, shrews, most viverrids). Their diets and fairly small body sizes probably constrain treeshrews to independent foraging. Within this paradigm, they could have a variety of social organizations. But they do not. As far as we know, all are both territorial and mostly monogamous.

The territorial social system of treeshrews sequesters food resources for territory holders (among other probable functions). Territory size, as I have argued above (chap. 8), is most likely determined by the foraging needs and ranges of reproductive females and their predispersal offspring. Support for this hypothesis comes from the studies of *Tupaia glis* in West Malaysia and Singapore (Kawamichi and Kawamichi 1979; Langham 1982), where that species occurred in very dense populations of tiny territories. The reproductive output of females in these dense populations remained similar or below that which I recorded for females in Sabah, implying that 0.9 ha *T. glis* territories on Singapore provided nutrients for similar breeding efforts as 7.0 ha territories of *T. longipes* in Sabah. This is amazing, and seems to show that social forces strongly constrain territories to minimal sizes that can only support breeding during part of the year.

The actual territory size (and consequent population density) thus seems to be related ultimately to the invertebrate-foraging mode of the species in question but proximately controlled by local food abundance and social interactions. It would be extraordinarily interesting to know the mechanism by which territory sizes are kept in the window below the size needed for year-round breeding and above that needed for subsistence and some breeding. Presumably a trade-off is achieved between the time/energy budget for territorial defense and that of foraging time. Kawamichi and Kawamichi (1979) saw large numbers of aggressive territorial chases within their dense study population, whereas such chases were rare in the rarefied populations I observed. The mechanisms of territory-size control by intraspecific competition could be studied experimentally by artificial manipulation of the food supply with removal and provisioning (as in Sherman and Eason 1998).

If food is just sufficient during lean times of year, then monogamy and pair-held territories may be the optimal social system, as each sex could prevent encroachment by same-sex rivals but would not need to expend energy on opposite-sex ones. The total lack of sexual dimorphism in body size of treeshrews argues that monogamy is the norm and perhaps that resources are insufficient for dimorphism. The role of the male in keeping other mouths off his territory so that more resources are available to his growing young, and to his female for reproduction, would be a large contribution of parental investment, even if he never directly interacts with his offspring.

DIET, HABITAT DIFFERENCES, AND CONSERVATION OF TREESHREWS

All treeshrews are forest creatures. What will happen to them as natural forests are altered by man? In chapter 4 we saw that each *Tupaia* species has a different ability to persist in logged forests and in tree plantations (Stuebing and Gasis 1989). With the picture of how each species uses the environment, we can now generate hypotheses to explain these differences and predict the future of treeshrews in altered habitats. To summarize the data (shown in table 4.4): (1) only *T. minor* is absent from logged forest, but it is present in tree plantations; (2) *T. tana* is present in plantations with dense understory but absent from plantations with open understory; (3) *T. longipes* is present everywhere and actually has increased densities in open understory plantations; and (4) *T. gracilis* is absent in all plantations but present in logged forest.

Because it is found in gardens and plantations, it is curious that *T. minor* is the only *Tupaia* adversely affected by selective logging. This may be due directly to severe disruption of the forest canopy, which is the foraging substrate for this species. Tree plantations usually have intact, uniform canopies, where lesser treeshrews can travel from tree to tree, but selective logging for dipterocarps destroys or damages up to 73 percent of the trees (Johns 1992). Logging may reduce the canopy beyond the point where lesser treeshrews can persist in the space of their usual home ranges.

Large treeshrews feed on earthworms and invertebrates in decomposing wood, on and under litter. Where the understory is dense, the ground surface litter layer is shady and moist, and arthropods are abundant near the surface; but in plantations where there is little ground cover, the surface is drier, and arthropods and worms may descend deeper into the soil and be absent near the surface. Moreover, such plantations may also lack decomposing wood and litter, as they are burned before planting. The total absence of slender treeshrews from plantations may be partly due to the same factors, but in addition, in primary forest this was the rarest species, with enormous home ranges and movements. A slender treeshrew might not be able to expand its daily foraging range any

Synthesis

farther if its resource density were any less than that found in undisturbed forest. This would eliminate slender treeshrews from all marginal habitats, and it is suggestive that this species likewise appeared to drop out at the lowest elevation on the elevational gradient of Mount Kinabalu. In contrast, *T. longipes* had the broadest habitat tolerance of all tupaiids, despite its huge territory size and foraging range in primary forest (i.e., relative rarity). Its densities even tripled in one exotic tree plantation, and I speculate that its major invertebrate resource, ants, may increase in secondary or altered vegetation, while undergrowth caterpillars, the specialty of *T. gracilis*, may decrease (more ants might even reduce the caterpillars). I thus conjecture that the different invertebrate foraging modes underlie the differing abilities of treeshrew species to occupy human-altered habitats.

In this context, it is noteworthy that the *T. glis* group of closely similar species, including *T. longipes*, have by far the widest geographic distribution of any treeshrews, throughout vast numbers of island and mainland habitats and regions as far as China (Corbet and Hill 1992). Lesser treeshrews have a large but smaller range to Thailand, large treeshrews range only to Sumatra, and slender treeshrews are endemic to Borneo. The extent of the biogeographic range is thus exactly mirrored in the number of types of exotic tree plots in which each species was found in a Sabah tree plantation. This may be only an amusing coincidence, but it could reflect the underlying ecological breadth of each taxon. In terms of their future conservation outlook, one can predict that the species are threatened by forest destruction in exactly the same order and pattern, with *T. gracilis* the most at risk and *T. longipes* the least. For pentail treeshrews, there are no pertinent data, but their large geographic range suggests broad habitat tolerance. For this species only, nest sites could be a limiting resource. All of the treeshrews will need some kind of forest to survive.

THE ECOLOGICAL PLACE OF TREESHREWS AMONG MAMMALS: WHAT IS A TREESHREW?

Mammalian orders group together families, genera, and species united by features that confer on them a distinct ecological place. Members of families are more tightly grouped into a narrower ecomorphological set than orders; and genera, more tightly still. If the treeshrews are in an order by themselves (Scandentia), we can ask whether the order or its subfamilies have any distinctive ecological or ecomorphological features.

PTILOCERCINAE

The nocturnal pentail treeshrews, Ptilocercinae, are in activity similar to the majority of small mammals. Diminutive nocturnal arboreal species are found among marsupials, insectivores, tenrecs, primates, and many families of rodents. Soft, woolly, pale gray fur, large mobile ears, long vibrissae, and the white tail flag are characters of pentails that are linked to nocturnal behavior and convergent with other nocturnal mammals but that are absent in *Tupaia*. The brilliant, pure white eyeshine of pentails is probably a unique character. As they are not arboreal leapers, but travel from tree to tree on connected paths, pentails should not require particularly good vision for locomotion. Their stunning eyeshine may indicate that vision is important for prey capture and/or that their exposed foraging substrate makes them unusually susceptible to predators that can be visually detected and avoided.

A survey of insectivory in arboreal mammals shows that although perhaps half of all species of nonbat mammals eat invertebrates taken from substrates above the ground, almost no mammal species except bats eat only invertebrates taken from arboreal substrates (Emmons n.d.), implying that pure arboreal insectivory is an unsuccessful strategy for mammals. About eleven small nocturnal mammals may have this feeding habit, but these are poorly known, and with one exception, all may eat some fruit and/or take terrestrial prey. The list includes pentail treeshrews, one mouse (*Dendroprionomys*), pygmy anteaters (*Cyclopes didactylus*), four species of tarsiers (*Tarsius* spp.), possibly scaly-tailed flying squirrels (*Zenkerella insignis*), and four species of trioks (*Dactylopsila* spp.). Tarsiers probably do not belong on the list, because they pounce onto prey on the ground, so they are not strictly arboreal foragers (P. Wright pers. com.), and trioks are said to eat some fruit; so possibly this list includes only two species, only one of them (*C. didactylus*) certain. Pentails are entirely arboreal foragers; if they turn out to feed almost wholly on invertebrates, then their ecological role seems to be almost unique among all living mammals, a quite surprising finding. If their diet turns out to include significant fruit, then they are in the company of scores of species.

However, the foraging substrate of pentails on trunks and stems is rare, perhaps shared by some Australian marsupials (Emmons n.d.). In Gabon I twice observed *Zenkerella* feeding nocturnally on termites on a tree trunk. Certainly no other mammal in its habitat in Sabah has the foraging characteristics of pentails (pers. obs.), which include speedy nocturnal gleaning of trunks and stems and extensive, rapid movement. Pen-

Synthesis

tails thus may have an activity period, diet, and foraging site combination that is either unique or shared by only a few rare mammals whose ecology is unknown, such as the African anomalurid *Zenkerella*, some Madagascar tenrecs of the genus *Microgale*, or the Neotropical mouse opossum *Glironia* (which I have seen move in the same way). By day, a number of tropical squirrel species, on all continents, glean insects from trunks, as do a few small primates, but they do this as a supplement to diets that include chiefly plant material (pers. obs.). Both arboreal insectivory and bark foraging on trunks are provinces into which specialist lizards and birds, but few mammals, have widely radiated. I believe that this may be because rapid movement between many trees is necessary for success, and for homeotherms this requires specialized locomotion, such as gliding, flying, or vertical cling-and-leaping. Running on wide, vertical trunks requires specialized morphology. Pentails have an extraordinarily strong finger grip with which they can drive the claws into bark, and when foraging they move with extraordinary speed.

TUPAIINAE

An outstanding ecological feature of Tupaiinae is obligate diurnality. Among all the living families of small mammals, only the Macroscelididae (elephant shrews, 15 species), Callithrichidae (marmosets and tamarins, 26–55 species, depending on taxonomy), Ochotonidae (pikas, 26 species) and Sciuridae (squirrels, 273 species) are chiefly diurnal (all but the flying squirrels). Other, largely nocturnal, taxa have some diurnal species, or species active both night and day, but almost all of these are cryptic animals that are largely active hidden under dense grass (voles and mice) or under the forest floor debris or moss (shrews, Madagascar *Nesomys* spp. rats), or small species active under dense forest understory or treefalls (short-tailed opossums, *Monodelphis* spp.; back-striped mice, *Hybomys* spp.) or close to safe refuges (pikas). Elephant shrews and pikas are all terrestrial (not scansorial), and pikas are grazing herbivores, so ecologically they have little in common with treeshrews. Other diurnal mammal families have larger-bodied species: only a single monkey is as small as a treeshrew (pygmy marmoset, *Cebuella pygmaea*). Thus only treeshrews and squirrels include conspicuous small, diurnal species that are arboreal and scansorial/terrestrial. The similarity in size, activity, and substrate use between tree squirrels and treeshrews is probably at the root of their uncanny physical resemblance. Are treeshrews and squirrels, or elephant shrews, also alike in other ecological features?

Worldwide, the tree squirrels individually and as a group can be characterized as feeders on tree seeds (mast) of all types. They are supremely efficient at feeding on hard, well-protected nuts, such as those of palms, and also on seeds protected by chemical defenses (Emmons 1975). They are seed predators that exploit the most energy-rich parts of fruits. The fruit-eating behavior of *Tupaia* differs sharply from that of squirrels. Treeshrews feed mainly on the soft, edible fruit parts "designed" as rewards for dispersers and eject large, hard seeds before consumption. My observations in Sabah suggest that treeshrews and squirrels feed on different fruit species, as well as on different parts of the fruit. Only one squirrel (*Callosciurus prevosti*) largely overlaps treeshrews in fruit preference, and this squirrel is a large, high-canopy species that directly competes for fruit only with *T. minor*, and which was never seen insect foraging. Treeshrews lack both the dentition to open hard nuts and the digestive apparatus to process much plant material. They profit from the abundant energy, and likely calcium, available in fruit crops by a batlike fruit-feeding pattern apparently geared to rapid digestion of readily assimilated nutrients during high-speed passage. This pattern may be the obligate result of their simple (primitive?) digestive tract morphology, but treeshrews are the only mammals other than bats that are known to feed in this way.

Arboreality within a family or genus in squirrels and other omnivorous mammals is associated with decreased insectivory, and terrestriality is linked to increased insectivory (Emmons 1980, 1995, n.d.). Treeshrews fit this pattern, in that most are terrestrial. Because it is arboreal, *T. minor* can be predicted to be more frugivorous than the other species.

The tupaiines can be classed as "generalized" insectivores. They all travel widely in search of dispersed, miscellaneous arthropods, but each species concentrates on particular sites and prey. In small insectivorous mammals, just as in birds, ecological separation between similar species seems to be most readily achieved by specialization to foraging for particular prey. This axis is evidently sufficient to permit the coexistence of arrays of congeners. In separating ecologically by insect-feeding guild more than by other parameters, treeshrews are similar to elephant shrews (Rathbun 1979), true shrews (Churchfield 1990), prosimians (Charles-Dominique 1971), monkeys (Terborgh 1983), or squirrels (Emmons 1980). The guilds represented in arrays of congeners in these taxa are often similar to those seen in treeshrews (ant eating, foliage searching, bark searching, height separation). This suggests that the invertebrate realm lends itself to being divided most easily in particular ways.

For example, in their manner of probing the leaf litter with the nose, *T. tana* are quite similar to golden-rumped elephant shrews (*Rhynchocyon chrysopygus*), which also feed on earthworms, among other prey. Other elephant shrews eat termites like *T. longipes* (*Elephantulus rufescens*), or are surface gleaners, like other tupaiids (*Petrodromus tetradactylus;* Rathbun 1979). Like treeshrews, elephant shrews are active for half of the twenty-four-hour day, implying that this type of general invertebrate gleaning requires long foraging times in both families. Certain elephant shrews are nearly entirely insectivorous (95–100%; Rathbun 1979), and they are cursorial, not scansorial, so they are in some ecological aspects quite unlike treeshrews.

The social systems of elephant shrews and treeshrews are similar (solitary animals on pair territories); in addition, they also leave their twin young, which are cursorial, in a separate shelter, and the mother has a type of absentee nursing. They thus share many characteristics with treeshrews, including diurnal activity (in the larger species), insect-foraging patterns, social system, and part of the reproductive pattern. However, they differ in their lack of frugivory or arboreality and in having precocious young, among other things, and thus they are not close ecological equivalents. What we know of both subfamilies of treeshrews thus far shows them to be distinct from other mammals in the combination of characteristics of their basic ecology, although no single feature is unique to them.

PTILOCERCUS AND *TUPAIA*: LIFESTYLE DIFFERENCES

Pentail treeshrews differ so much in morphology from other living treeshrews that in every complex character set examined they are placed on a separate lineage from other genera, at the level of a different subfamily (Luckett 1980). It is therefore instructive to see whether these morphological differences among the subfamilies are mirrored by differences in their behavioral ecology. The small amount of field data that we were able to collect on a single group of pentails indicate that this is indeed the case.

It was formerly known that pentails were the only nocturnal treeshrews and that several could be found in a nest together (Gould 1978; Lim 1967). The social organization of pentails is based on what appears to be a tightly coherent family group that lives permanently together in a single den site. Young pentails share the den with the mother, and they seem to remain with her until they reach adult size, after the birth of the next litter. I believe that pentails do not have the "absentee" parental care system, because tiny, newly emergent young already shared the mother's den. Young appeared to forage in separate "beats" within the maternal home range while they continued to share the communal nest. Pentails lead a nest-centered life, with foraging trajectories always looping out from the nest and back to it. Group members leave and return to the nest close together in time and interact on and near the den tree. Pentails therefore seem to be highly social animals that spend the twelve hours of daylight sleeping with their family or group but forage solitarily.

Previously, it was thought that *Tupaia* also probably lived in monogamous pairs that shared a nest (Martin 1968). However, our radio-tracking showed that male-female pairs that shared a territory never slept together in the same nest. At least three of the *Tupaia* species were sometimes polygynous, and one, at least, could be polyandrous, according to their territorial deployment and the behavior of males. Monogamy is thus only facultative, not obligate or fundamental to the social system, although it was the most usual condition. The young of *T. tana* that we followed from near birth spent only the one or two nights after their emergence from the nursery nest with their mother. Neither adult pairs nor mother and young therefore nested together, and in general *Tupaia* could be thought of as largely solitary-living animals that often met and interacted with their territory mates (usually at least daily) during otherwise solitary foraging rounds. However, some species (*T. minor, T. montana*) show more within-pair cohesion than others (*T. gracilis, T. tana*), and pairs sometimes forage together. In lesser treeshrews, pair foraging in exposed arboreal trajectories may help in the detection of raptors. Tupaiines therefore may sometimes forage together but never nest together, while just the opposite is the case in Ptilicercines.

PHYLOGENY, ARBOREALITY, AND TERRESTRIALITY

Morphologically, *Ptilocercus* falls on cladograms as the least derived, most primitive genus of treeshrew. For example, in a phylogeny constructed from nine diverse characters that show variation within treeshrew genera (Luckett 1980: fig. 7), *Ptilocercus* has the primitive state for all nine. Of seven foot-bone characters in which *Ptilocercus* and Tupaiines differ, *Ptilocercus* has the primitive state for six and *Tupaia* for one (Novacek 1980). The same situation holds for all other character sets, such as teeth (Butler 1980).

Synthesis

Jenkins (1974) has argued that there is no sharp dichotomy between arboreality and terrestriality for small forest mammals such as treeshrews, because the forest floor is not an even surface but a cluttered jumble of fallen logs, branches, roots, and irregularities, where versatile locomotion including climbing ability is advantageous, so that early mammals were likely to have been neither arboreal nor terrestrial but scansorial, as are modern didelphid marsupials, *Rattus*, and tupaiids.

> Accordingly, the question of an arboreal or terrestrial ancestry for mammals no longer is relevant. The locomotor niche which ancestral mammals exploited undoubtedly included both "terrestrial" and "arboreal" surfaces. Ancestral primates very likely occupied forest habitats in the manner of some living treeshrews. The adaptive innovation of ancestral primates was therefore not the invasion of the arboreal habitat, but their successful restriction to it. (Jenkins 1974: 112)

This scenario is reasonable and likely valid for primates, but I argue that for treeshrews, diversification probably took place in the contrary manner: by descent from the trees into terrestrial foraging modes. My hypothesis is as follows:

1. The most primitive living treeshrew (pentail) is strictly arboreal, so much so that it is a klutz on the ground (Gould 1978). It is not in any way "scansorial." At least one *Tupaia* species is also strongly arboreally adapted, so at least two species are restricted to the trees. All treeshrews have morphological adaptations for true arboreality (rotating ankles; Szalay and Drawhorn 1980), and some of the most terrestrial species nest in trees (*T. tana, T. gracilis*). This argues for arboreal antecedents, as tree-descending ankles and nesting are unlikely to have evolved without truly arboreal lifestyles.

2. Among approximately 3,500 living nonvolant mammal species, only five to seven species, at most, feed entirely on arboreal invertebrates. Therefore purely arboreal insectivory is a limited, practically nonexistent niche. In contrast, hundreds of terrestrial and scansorial species are purely or almost purely insectivorous. Strong insectivory is the domain of terrestrial mammals (Emmons 1995); in some places sixteen true shrew species can coexist (Ray 1996).

3. To radiate arboreally, as have arboreal rats, primates, squirrels, palm civets, Australian marsupials, and so forth, an enhanced morphological/physiological ability to use primary productivity of the canopy

(fruit, leaves) is required—especially a more complex digestive tract. The fruit-feeding mode of treeshrews is rudimentary and as such does not offer much opportunity for dietary diversification.

4. With the diet and anatomy that they have, treeshrews therefore only seem likely to diversify by adding terrestrial species with different invertebrate foraging modes. There is never more than one arboreal *Tupaia* in any community; but as shown in this book, a number of syntopic terrestrial species co-occur and exploit the invertebrate prey base in different ways. The greatest radiation is found among the terrestrial forms, leaving the primitive arboreal forms behind in the trees.

As an evolutionary experiment, modern members of the Tupaiinae seem to be improbable models or precursors for primates. They show signs of having radiated by differentiating into terrestrial faunivorous guilds, which seems unlikely to be a scenario from which early primatelike species would develop. Moreover, they have evolved in the contrary sense in terms of one of the hallmarks of primates, parental care, with even less care than the most primitive living mammals. That said, the other branch of the treeshrews, the pentails, with their cohesive family groups and nocturnal, arboreal lives, seem to be an excellent model of a species antecedent to prosimian lifestyles. An increase in frugivory—a likely development in an arboreal mammal—would easily lead to evolution in a primatelike direction.

The importance of specialization to frugivory for primate radiation and evolution is clearly shown by the reproductive limitations of wild *Tupaia*, which although chiefly faunivorous, appear to depend on fruit for the extra calories needed for reproduction. Without a turn toward the more complex digestive tracts, teeth, and hands needed for better processing of arboreal fruits, primate ancestors may have remained locked into an energetically limited treeshrewlike ecology with little scope for radiation. The all-consuming quest for arthropods in the faunivorous diet likewise precludes devotion of much time to social interaction, except, as in pentails, in the nest during inactivity.

As a model of how Miocene mammals might have lived, scampering among the feet of giant reptiles, *Tupaia* may be far more interesting than it is as a model for early primates. Tupaiine treeshrews, with their generalized body plans, engage in a variety of ecological roles that permit the coexistence of an array of sympatric species. Early mammals should have likewise been able to exploit many invertebrate-eating guilds without great morphological modification.

CONCLUSIONS

The Scandentia is a small order of small, inconspicuous species that live in the forests of Asia. One subfamily, Tupaiinae, includes a radiation of perhaps twenty species, whereas the more primitive Ptilocercinae has only one living member. Both the parental care system and lifestyle of the Tupaiinae are unique and more distinctive from those of other mammals than are those of *Ptilocercus*. I believe tupaiines to be derived from more arboreal earlier mammals with a more standard behavioral ecology, such as *Ptilocercus*. The small number of living tupaiid species and the depauperate fossil record suggest that their characteristics never provided much potential for evolutionary diversification, yet a few living species are common and highly successful, in terms of their numbers and the wide array of environments that they occupy. They are therefore an evolutionary success, able to hold their own amid more populous orders. Tupaiid absentee maternal care is an extremely specialized and effective system, but, like other specializations, it strongly limits its possessors. The absentee maternal nursing system "works," but it provides little flexibility in litter size or maternal behavior, and it would not seem likely to offer much potential for evolving into other strategies needed for treeshrews to succeed in very cold climates or in places where survivorship of young was much lower and litters of two young would not suffice. That only a few species persist in continental Asia, while quite large arrays of sympatric species coexist on Borneo, suggests that treeshrews may be at a disadvantage under more intense faunal competition on the mainland. But that they do persist on the mainland, and often in very high numbers, is also evidence that they have a unique ecological place that removes them from intense competition with members of more advanced and speciose mammalian families.

APPENDIX I

Methods

TRAPPING

Treeshrews were trapped with locally purchased cage traps of small poultry wire mesh, baited with a slice of ripe plantain or banana. Before a trapping session, traps were fixed open and prebaited with several hunks of banana once or twice at two- to three-day intervals. Prebaiting saved time by training the animals to come to trap sites, and I hoped it would also reduce trap-shyness by habituating treeshrews to being rewarded in traps without the stress of capture. Prebaiting brilliantly succeeded in the first goal, because most animals were captured the first day traps were set. There is no control data by which to judge the second objective. Toward the end of the study, we also used mixed concentrated banana and strawberry essence on a cotton swab placed in the trap, as an attractant (A. Rabinowitz pers. com.) along with banana. On the high plot at Poring we tried baiting with a mixture of peanut butter, oatmeal, fermented prawns (balachan), and dried beef shreds, in a fruitless attempt to capture the other montane treeshrew, *Dendrogale melanura*. In 1989 and 1990 traps were covered with fresh leaves to shelter the animals from sun and rain, but beginning in the rainy season in 1990 a plastic bag was tied with string over the top of each trap.

In 1989 and 1990 I set traps at nightfall (1700 h–1800 h) and left them open at night. In 1991 a population explosion of rats filled 50 percent of the traps each night. Thereafter, to avoid interference by rodents, for the first trapping day of the session I set traps before dawn, from 0430 to 0600 h, but then left them open the following night to monitor the rat populations. In 1989 we checked traps twice during the day, beginning at 0900 and 1700; but in 1990–91 I usually visited traps three times, beginning about 0800, 1400, and 1630 h. The octagonal wire mesh of the traps was large enough for treeshrews to poke their muzzles through, which could injure the skin on the rostrum. Incidence of dam-

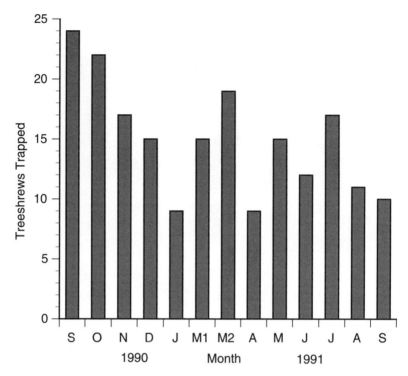

Fig. A-I.1. Total number of treeshrews trapped during each monthly trapping session at Danum Valley, September 1990 to September 1991. The trapping sessions were not all equivalent; the September 1990 session was longer, and that in January was curtailed by elephants.

age from this injury increased with the amount of time animals stayed in the traps, and with the number of times they were recaptured during the same session. To minimize this problem, and generally reduce the risk of accidents, I trapped for only two days each session, after an initial four- or five-day session at each study site to capture as many individuals as possible. Two days were not long enough to capture all treeshrews in the area.

At the lowland, Eastern Ridge site at Poring (1989), traps were placed at 20 m intervals, on the nearest likely sites, in a 6 × 8 grid, with a 2 × 10 extension, designed to cover the available flat area of the ridge top. The area of the ridge top within the trails was 6 ha, that inside the trap array was 3 ha. The 1,000 m elevation, Langanan Ridge site, where *T. montana* was studied, had a wedge-shaped grid of 9 × 10 traps, tapering to 4, which again filled the flat space of about 4 ha on the ridge top. The home ranges of most species turned out to be so large that a grid arrangement on a practical scale for acquiring home range data from trapping was of little use (I had originally hoped to supplement radio-tracking by trapping), so at Danum Valley in 1990 I abandoned the grid pattern and

Appendix I

arranged traps at 25 m intervals in two 500 m lines 100 m apart. At all times traps were placed on the nearest likely trapping spot to each station (fallen log, dense brush, treefall). At twenty-three places on the Eastern Ridge grid at Poring and at seven places on each line at Danum Valley, additional traps were tied up in vine tangles or on branches about 2–3 m above ground. These were put in likely spots, without reference to the other trap markers.

In 1989 we trapped the Eastern Ridge grid at Poring in the first weeks of April, May, and finally in mid-August; and Langanan Ridge at the beginning and end of July. We also ran a small altitudinal transect with 30 traps from 880 m to 1,000 m on Langanan Ridge for three days in June 1989. At Danum Valley we trapped each month toward the end of the month from September 1990 to September 1991, except February was missed, and March was trapped twice, at its beginning and end. In the January trapping period, a group of elephants went through the study area, systematically destroying man-made objects (rain gauges, signs, trail markers, etc.). They stomped on seven baited traps but interestingly did not step on one that contained a treeshrew, while flattening the two on each side (trapping in January was curtailed, and this month is excluded from most data). Trapping success was very high. At Danum the number of individual treeshrews captured per month ranged from nine to twenty-four (fig. A-I.1).

HANDLING TREESHREWS

Trapped animals were transferred to cloth bags, weighed, measured, and ear-tagged with numbered monel self-piercing tags. In addition, the tail hair was cut in a unique pattern for each individual, to aid visual recognition in the field (Kawamichi and Kawamichi 1979). The first treeshrew captured, a female *T. tana*, died of stress shock while being held gently (a problem also common in many tropical squirrel species). Thereafter, I anesthetized all treeshrews by injecting them with ketamine hydrochloride, through the holding bag, before initial handling and fitting of radio-collars. No other mortality ever occurred during handling. After handling, anesthetized animals were allowed to recover in a cloth bag and released on the next round of trap checking. If a treeshrew to be collared was caught at nightfall, I kept it overnight in the safety of my room and released it in the morning. Later in the study, I found that neither *T. longipes, T. minor*, nor *T. montana* were susceptible to shock, and these species were handled without anesthesia, except when collars were fitted. I was reluctant to anesthetize treeshrews repeatedly, so when animals were recaptured, I did not anesthetize them but usually weighed them in the bag, examined the reproductive condition of females, and released them without further handling.

PROBLEMS ENCOUNTERED

To help those who study treeshrews in the future, I briefly sketch the problems or failures with our methodology. The traps themselves caused a few injuries: one treeshrew died when the door fell on its neck; one caught its leg in the wire and broke it; two tore their skin badly on the protruding wire of the spring; and one caught its radio-collar in the bait hook. All injured treeshrews were immediately released without treatment or handling, and all survived the injuries and were recaptured with healed wounds in the following months, including the broken leg. In my experience from trapping many mammal species, injured animals survive far better when immediately released and left untreated in the wild than when sentimental humans attempt to "hospitalize" them. We made 420 treeshrew captures during the study, so the casualty rate from the trapping procedures was extremely low.

The use of ear tags was not a success, because many were lost. When treeshrews that had lost ear tags were recaptured, I permanently marked them by clipping a terminal toe phalange as is the standard method in mammalogy. This also was not successful because several individuals, especially *T. longipes*, then became trap-shy. One of these regularly came to bait sites, where he was repeatedly seen (he had a radio-collar), but he was never caught again during the remaining six months of the study. About a half dozen animals, mostly young, were recaptured without ear tags after several months, with tail hairs grown out, so that I lost their identities and thus lost valuable information on longevity and the dispersal of young. In future, I would use electronic numbers on transponders for permanent marking.

The radio-collars seemed to cause few problems in large species. One female *T. tana* whose collar was tight was recaptured in a state of shock, so her collar was removed. She was recaptured the next month in perfect condition and recollared with no problem. One male *T. tana* was found dead sixteen days after collaring. He may also have been stressed by the collar, but he was in good condition, robust and glossy, with no evident injury or external signs of stress. A number of collars slipped off because they were too loose.

Animals should not be released before full recovery from anesthesia. All three small species, but no large ones, had a tendency to get a forefoot caught between the collar and the neck. This happened especially when the animals were first collared and coming out of anesthesia (*Ptilocercus* particularly), when they probably tried to push the collar off. There may be no way to avoid this problem, but collars should be carefully fitted, and not too loose. Radios were always removed at the end of the study or tracking session, but several animals were recollared, or were trap-shy, and wore radios for six to twelve months without evidence of any injury from the collars.

The most trouble-free radio-collars for large species were those with a brass band enclosed in shrink-tubing that formed a closed loop antenna that was bolted on (Wildlife Materials, Inc.). I had an extra hole drilled on each side of the tuned diameter center so that fit could be adjusted. Whip antennas were tried, but these always broke off rapidly. Teflon fabric collars frayed and broke.

RADIO-TRACKING

The study areas were gridded with trails every 100 m and marked with meter marks every 20 m (Poring) or 25 m (Danum). Radio-tagged treeshrews were fol-

Appendix I

lowed on foot on the study-area trail networks, and signals were triangulated from three trail marks approximately every 20 minutes. If the animal was moving rapidly, it was followed continuously, so that it would not escape from range. During a sample, the treeshrew was followed from before it emerged from its nest until after it retired.

Telemetry data was hand-plotted for best accuracy (some bad triangulations must always be discarded, and computers are not good at choosing the right data to reject). Distances moved and areas of ranges were measured from plotted maps with a digitizing tablet and DesignCad ® software.

Time only allowed each person/team to do intensive radio-tracking of 6 to 7 individual treeshrews per month for three-day samples, or 4 to 5 animals for five-day samples. The chief drawback of the continuous following method is that few animals can be tracked in any month, and the necessity to follow treeshrews of both sexes of five species meant that often an individual was only followed for one sample. I tried to radio-collar all resident adults of every species except *T. minor*, where my object was to collar only contiguous females or both members of a pair. I gave priority to putting radios on males whose females had working radios, so that interactions could be monitored during tracking, but sometimes the timing of when individuals were captured, when radios functioned, and when we could schedule tracking did not work out, so that tracking of contiguous or overlapping individuals was staggered in time.

APPENDIX II

Fruit Species Collected at Danum Valley

Fruit species collected on the Danum Valley study area during 1990–91 on monthly fruit transects. Species identified and chiefly collected by Elaine Campbell Gasis. Cap = capsule.

Species	Form	Fruit L (mm)	Fruit W (mm)	Fruit Type
Alangiaceae				
Alangium ebenaceum	tree	2	1.5	drupe
Alangium griffithii	tree	2.2	1.3	drupe
Anacardiaceae				
Buchanania sessilifolia	tree	0.8	0	drupe
Anonaceae				
Artabotrys roseus	liana	4	4	berry 2 seed
Polyalthia cauliflora	tree	1.6	1.6	berry 2 seed
Polyalthia rumphii	tree	1	0.8	berry 1 seed
Polyalthia sumatrana	tree	2.8	1.2	berry 1 seed
Polyalthia sp. B		1	0.7	berry 1 seed
Polyalthia sp. C		1	1	4–5 seeds
Uvaria grandiflora	liana	2.7	1.2	
Uvaria javana				
Anonac. sp. A		2.9	2.4	multiple
Anonac. sp. B ? seed only				
Anonac. sp. C				
Apocynaceae				
Willughbeia sp. A	liana	6	5.5	berry 10–20
Wrightia sp. A	liana			
Burseraceae				
Canarium decumanum	tree	5	4	drupe

Appendix II

Species	Form	Fruit L (mm)	Fruit W (mm)	Fruit Type
Canarium denticulatum	tree	3	1.5	drupe
Celastraceae				
Kokoona reflexa	tree			
Celastrac. sp. B		17	3	pod
Combretaceae				
Combretum nigrescens	liana	3.5	2.5	4 wing
Connaraceae				
Agaelaea borneensis	liana	2	1	caps aril
Convolvulaceae				
Convol. sp. A	liana	0.6	0.3	capsule
Convol. sp. B	liana	1.6	0.8	capsule
Convol. sp. C	liana	2.8	1	
Convol. sp. D	liana			
Dilleniaceae				
Dillenia excelsa	tree			capsule
Tetracera akara	liana	1	3	caps aril
Dipterocarpaceae				
Parashorea malaanonan				wing
Shorea johorensis				wing
Shorea leprosula				wing
Shorea parvifolia				wing
Shorea paucifolia				wing
Ebenaceae				
Diospyros macrophylla	tree	8.5	5.8	berry
Diospyros sp. A	tree	1.1	1.2	
Diospyros sp. B	tree			
Euphorbiaceae				
Antidesma neurocarpum	tree	0.8	0.6	berry?
Antidesma sp. A	shrub?			
Aporusa acuminatissima	tree	4.5	3.5	berry
Aporusa nitida	tree	4.5	3	schizocarp
Baccaurea stipitata	tree	2	1.5	berry 2 seed
Bridelia sp. A	tree	1.3	1.2	drupe
Glochidion sp. A	tree			
Macaranga conifera	treelet	0.8	0.4	berry 2 seed
Mallotus lackeyii	tree	0.6	0.4	berry?
Mallotus sp. A	tree	1.5	1	cap 3 loc
Mallotus sp. B *Spathiostemon?*	tree	1.6	1.6	cap 3 loc
Mallotus sp. C *Spathiostemon* or *cleistanthus*	tree	1.4	0.8	cap 3 loc
Neoscortechinia forbesii	tree	3.5	2.3	schizocarp 4
Fagacae				
Castanopsis costata	tree	4	4	nut hard
Castanopsis sp. D *evansii*?	tree	8	6	nut hard
Castanopsis hypophoenicea	tree	9	7	nut hard
Castanopsis sp. E	tree	4	3	nut
Castanopsis sp. C	tree	3	2.3	seed only

Appendix II

Species	Form	Fruit L (mm)	Fruit W (mm)	Fruit Type
Lithocarpus ewyckii	tree	2	1.5	acorn
Lithocarpus gracilis	tree	1.7	1.8	acorn
Lithocarpus leptogyne	tree	2.5	2	acorn
Lithocarpus sp. A	tree	1.5	2	acorn
Lithocarpus sp. E	tree	2.2	1.8	acorn
Flacourtiaceae				
Flacourt. sp. A?		3.1	2.7	drupe
Gonystylaceae				
Gonystylus keithii	tree			
Guttiferae				
Callophyllum sp. A	tree			
Garcinia forbesii	treelet	3.8	3	berrylike aril
Garcinia parvifolia	tree	2.5	2.5	berrylike aril
Juglandaceae				
Engelhardia serrata	tree	4	0.8	wings nut
Lauraceae				
Litsea sp. A?		3	1.8	
Lecythidaceae				
Barringtonia sarcostachys				
Barringtonia sp. A				
Leeaceae				
Leea aculeata	tree	2	1	berry
Leguminosae				
Bauhinia sylvanii	liana	13.5	5	pod 2 seed
Cesalpinia oppositifolia	Liana	8.5	4	pod 2 seed
Dialium indum	tree	2.7	2	pod edible pulp
Fordia splendidissima	tree			
Intsia palembanica				
Koompassia excelsa	tree	10	3.2	winged seed
Mucuna bifurcata	liana	8	5	pod 2 seed
Spatholobus sp. A	liana?	14.5	4	pod 1 seed
Leg. sp. H		8	3.2	pod woody
Leg. sp. K				
Melastomataceae				
Medinila sp. A	epiphyte	0.6	0.6	berry
Melastom. sp. A				
Meliaceae				
Aglaia sp. A	tree			
Aglaia sp. B	tree			
Chistocheton ceramicus	treelet	3.2	2.5	loc arils
Dysoxylon alliaceum	tree	3	2.5	locules
Dysoxylon cyrtobotryum	tree			
Dysoxylon sp. B	tree	4	3.5	locules?
Dysoxylon sp. C	tree	8.3	6	loc aril
Dysoxylon sp. D	tree			
Walsura sp. A	tree	10	7	loc 2 seeds
Meliac. sp. D	tree	6.8	6.5	loc 2 seeds

Appendix II

Species	Form	Fruit L (mm)	Fruit W (mm)	Fruit Type
Meliac. sp. E		4	3.8	
Meliac. sp. H	tree			
Meliac. sp. I	tree			
Menispermaceae				
Stephania sp. A seed only		1.8	1.5	seeds only
Stephania sp. B seed only		3.7	1.8	seed only
Moraceae				
Artocarpus nitidus				
Ficus xylophylla	tall tree	1.2	1.1	synconium
Ficus sp. A		2.5	1.8	synconium
Ficus sp. B		1.4	1.2	synconium
Ficus sp. C		0.8	0.7	synconium
Ficus sp. D				
Ficus sp. E				
Myrsinaceae				
Ardisia macrophylla	treelet	1.8	1	drupe
Myrtaceae				
Eugenia cerasiformis	tree	1.8	0.7	drupe
Eugenia fastigata	tree	1.9	1.5	drupe
Eugenia lineata	tree	0	0	drupe
Eugenia sp. A		2.2	1.5	drupe
Eugenia sp. B				
Eugenia sp. C				
Eugenia sp. D				
Olacacae				
Scorodocarpus borneensis	tree	5	3	drupe
Palmae				
Rattan sp. A	liana	1.4	0.7	
Passifloraceae				
Adenia macrophylla	liana	6.5	3.5	caps aril
Proteaceae				
Helicia robusta	tree	4.3	5	berry 3–4 nuts
Rhamnaceae				
Zizyphus borneensis	tree	2.7	2.2	drupe
Rubiaceae				
Uncaria sp. A	liana	4.5	0.3	pod achene
Urophyllum glabrum	liana			drupe
Rutaceae				
Luvunga sp. A	liana	1.7	0.7	drupe
Sapindaceae				
Dimocarpus longan	tree	1.5	1.5	drupe
Nephelium maingayi	tree	3.5	2.5	drupe
Nephelium rambutan-ake	tree	5	4.5	drupe
Paranephelium xestophyllum	tree	4	3	berry 3 loc
Pometia pinnata	tree			

Species	Form	Fruit L (mm)	Fruit W (mm)	Fruit Type
Sapotaceae				
Payena acuminata	treelet	3	0.8	drupe
Saxifragaceae				
Polysoma integrifolia		2.2	1	drupe
Solanaceae				
Solanum incanum	epiphyte	1.5	1.5	berry
Sterculeaceae				
Pterospermum stapfianum	treelet	5.5	1.5	pod wing seed
Tiliaceae				
Microcos antidesmifolia	tree	1.8	1.8	drupe
Microcos reticulata	tree	3	1.5	drupe
Pentace laxiflora	tree	1.7	1.2	5 wings
Urticaceae				
Poikilospermum sp. A	epiphyte	6	6	aggregate
Verbenaceae				
Callicarpa longifolia	shrub			
Congea tomentosa	liana			
Vitaceae				
Parthenocissus sp. A	liana	1.8	2	berry
Tetrastigma trifoliatum	liana	1.3	0.8	berry
Zingiberaceae				
Globba pendula				

APPENDIX III
Mammal Species Found on the Study Plots

List of mammals we identified on the study plots (X), or present in nearby areas (P). For Poring, only species we identified during the study are listed, as we cannot verify the continued presence of historic records. A mammal list for Danum Valley can be found in Anon. 1993.

Species	Mass (kg)	Poring Low	Poring High	Danum
Echinosorex gymnurus				X
Crocidura sp.				P
Crocidura sp.				P
Tupaia gracilis	0.080	X		X
Tupaia longipes	0.215	X	X	X
Tupaia minor	0.060	X	X	X
Tupaia montana	0.177		X	
Tupaia tana	0.230	X	X	X
Ptilocercus lowii	0.050			X
Cynocephalus variegatus	1.000	X		X
Pteropus vampyrus	1.000			P
Cynopterus brachyotis				P
Rhinolophus acuminatus				P
Rhinolophus borneensis				P
Rhinolophus sedulus				P
Rhinolophus trifoliatus				P
Hipposideros galeritus				P
Tarsius bancanus	0.110	X		X
Nycticebus coucang	3.500	X		X
Macaca fascicularis	4.800			X
Macaca nemestrina	6.500	X		X

Species	Mass (kg)	Poring Low	Poring High	Danum
Nasalis larvatus	16.000			P
Presbytis cristata	5.300			P
Presbytis hosei	6.500			P
Presbytis rubicunda	6.300	X	X	X
Hylobates muelleri	5.700			X
Pongo pygmaeus	30.000			X
Prionailurus bengalensis	5.000	X	X	X
Prionailurus planiceps				P
Pardofelis marmorata				X
Neofelis nebulosa	19.500	X		P
Herpestes brachyurus				X
Herpestes semitorquatus				X
Amblonyx cinereus				X
Lutra sumatrana				X
Lutrogale perspicillata				X
Martes flavigula		X	X	X
Mustela nudipes		X		X
Mydaus javanensis				X
Helarctos malayanus	55.000			P
Arctictis binturong	6.700			P
Arctogalidia trivirgata	2.200	X	X	X
Cynogale bennettii				P
Hemigalus derbyanus				X
Paradoxurus hermaphroditis	3.000	X		X
Prionodon linsang	0.700	P		
Viverra tangalunga				X
Manis javanica	6.000			X
Elephas maximus				X
Dicerorhinus sumatrensis				P
Sus barbatus		X	X	X
Tragulus javanicus		X		X
Tragulus napu				X
Cervus unicolor				X
Muntiacus atherodes				X
Muntiacus muntjak		X		X
Bos javanicus				P
Aeromys tephromelas	0.900			X
Aeromys thomasi	1.450			X
Callosciurus adamsi	0.130			P
Callosciurus notatus	0.185	X		
Callosciurus prevosti	0.380	X		X
Callosciurus baluensis			X	
Dremomys everetti			X	
Exilisciurus exilis	0.020	X		X
Exilisciurus whiteheadi		X	X	
Lariscus hosei		X		
Ratufa affinis	1.150	X	X	X
Petaurillus sp.	0.025	P		
Petaurista petaurista	2.000			X

Appendix III

Species	Mass (kg)	Poring Low	Poring High	Danum
Sundasciurus brookei		X	X	
Sundasciurus hippurus	0.310	X	X	X
Sundasciurus jentinki			X	
Sundasciurus lowii	0.450		X?	X
Haeromys cf. *margarettae*	0.010	X?		X
Lenothrix canus	0.160			X
Leopoldamys sabanus	0.550	X	X	X
Maxomys rajah		X		X
Maxomys surifer		X	X	X
Maxomys cf. *whiteheadi*		X	X	X
Niviventer cremoriventer	0.070	X		X
Pithecheirops otion	0.130			P
Sundamys muelleri				X
Hystrix brachyurus		P		P
Thecurus crassispinis				X
Trichys fasciculata		X		X

APPENDIX IV

Invertebrates in Treeshrew Diets

The following tables list the occurrence of items in treeshrew diets, and the taxa that could be identified. Number of scats in which item found (O), and total number of individuals identified in scats (I), except ants and termites, where the number in a scat is classed as: 1 = <10; 2 = 10–20; 3 = 20–30 or "many"; 4 = >30. Because not all items could be identified to family, items in family rows do not necessarily add up to the number in orders. Identifications are by Warren Steiner.

Food	P. lowii (12) O	I	T. minor (28) O	I	T. gracilis (86) O	I	T. montana (99) O	I	T. longipes (220) O	I	T. tana (514) O	I
Earthworm							3	3			33	29
Lepidoptera larva	1	1	2	3	20	34	3		5	5	20	6
Lepidoptera adult	1	1	1	1	3	3			2	2	6	33
Coleoptera larva							1	1	3	3	24	6
Elaterid								1				9
Scarabid												3
Carabid												3
Phengatidae												3
Coleoptera adult	1	1	3	6	4	4	11	21	17	24	35	62
Anthribidae										1		1
Carabidae								2				3
Cerambycidae						1		1		4		2
Chrysomelidae		1						1				
Coccinellid								1				
Curculionid				1				3		2		2
Endomychidae										1		1
Elateridae												1
Lucanidae												1
Ptiliidae												1
Scarabidae	2			1				1		3		18
Acanthocerinae												1
Melolonthine						1		3				7
Dynastininae								1				
Rutelinae												1
Scloytid				1						1		2

Food	P. lowii (12)		T. minor (28)		T. gracilis (86)		T. montana (99)		T. longipes (220)		T. tana (514)	
	O	I	O	I	O	I	O	I	O	I	O	I
Staphylinid	1											4
Tenebrionid		1								3		1
Alleculine										2		
Formicinae												
Larvae/eggs					18	20	14	26	41	104	54	86
Alates						1				5		1
Isoptera												
Alate									30	53	16	26
Hymenoptera										4		
Aculeate wasp					1	1	3	3	2	3	5	6
Ceraphronidae												
Ichnumonidae								1				1
Sphecidae						1				1		1
"Wasp"								1				
Orthoptera												
Acridid	1	2	4	14	16	18	11	17	7	7	47	102
Blattid				5		4		5				4
Gryllid				4		3		4		2		54
Gryllotalpid												20
Tettigoniid						1						2
Tetrigid?										1		10
Tree cricket		2								1		1
Hemiptera	1	1			2	2	3	3	10	15	14	14
Reduviid		1				1		3		13		7

	1	2	3	4	5	6	7	8
Coreid								4
Cydnid						2	4	6
Diptera			1					1
Tipulid			1					
Homoptera			2	1	1	1	1	1
Cicada					1			
Fulgorid			1					
Odonata			1	10	12	2	33	1
Aranae	1	3			4	2		51
Lycosid		1			1	1		24
Salticid					4		16	21
Diplopoda	1	1		3				5
Glomerida								1
Spirobolid?							17	19
Chilopoda				2	2			9
Scolopendrid					2		4	4
Uropygida							1	1
Phalangida							9	9
Scorpions							3	3
Decapoda							1	1
Crayfish								
Crab							2	2
Mollusca	1	1						
Lizard						1	1	1
Lizard/snake						1		

APPENDIX V

Consumers of Fruit Species

List of fruit species seen eaten by various mammals and birds during the study. Site: D = Danum; P = Poring. ? = treeshrew with radio nearby, but it is not certain that it ate the fruit.

Fruit	Site	Consumers
A. Fruits Eaten by Treeshrews		
Alangium ebenaceum, 13–29 July 1991	D	*T. gracilis; T. longipes; T. tana; Callosciurus prevosti; Sundasciurus lowii; Presbytis rubicunda; Macaca fasicularis; Hylobates muelleri; Pongo pygmaeus; Viverra tangalunga; Sus barbatus; Muntiacus atherodes; Tragulus napu*
Dialium indum, 3–29 December	D	*T. longipes; Hylobates muelleri; Presbytis rubicunda; Macaca fascicularis; Muntiacus muntjac; M. atherodes; Tragulus napu; Callosciurus prevosti*
Dimocarpus longan, 16–28 September	D	*T. longipes; T. tana; T. gracilis; Hemigalus derbyanus; Buceros rhinoceros*
Ficus, 14 April, II-7	P	*C. prevosti*; unidentified monkeys
Ficus, 13 May, VI-2	P	*T. longipes; T. gracilis*
Ficus, 15 May, C60	P	*T. minor; T longipes; T. gracilis; C. prevosti; S. lowii*
Ficus, 20 May, 3.175	P	*T. longipes*; birds

Appendix V

Fruit	Site	Consumers
Ficus, 25 May A660	P	*T. glis; C. prevosti*; many birds
Ficus, 25 May, C9	P	*T. minor; T. gracilis?*
Ficus, 25 May, C20–40	P	*T. minor; C. prevosti; S. brookei*; Barbets; many birds
Ficus, 5 August, A480	P	*T. gracilis; Calyptomena viridis; C. hosei; Megalaima henricci*; many pigeons
Ficus G., 14 June	D	*C. prevosti*; many birds
Lansium domesticum, 18 October	D	*T. longipes*
cf. *Madhuca* sp., 20 August	P	*T. tana; T. longipes*
Myrsinaceae, 11–14 April	P	*T. minor*; birds
Payena acuminata, 31 August	D	*T. tana*
Rafflesia keithii	P	*T. tana; Callosciurus notatus*
Sapindaceae, 31 July	P	*T. montana*
Parthenocissus sp. A, 8 October	D	*T. minor*
Sp. J *Polyosma integrifolia?* 18 October	D	*T. glis; T. tana*
Polyalthia sumatrana, 26 October–28 November	D	*T. tana; T. glis; Viverra tangalunga; Anorrhinus galeritus*
Ficus (tiny epiphyte, W1, N020), 27 March	D	*T. tana; T. minor; T. longipes*; one bird
Ficus (giant canopy, W2N225), 26 July	D	*T. gracilis; T. longipes; P. lowii?; T. napu*; many birds
Ficus (giant emergent, N1W250), 28 August 1991	D	*T. longipes*
Diospyros, 8 June 1991	D	*T. minor?; Buceros rhinoceros*

B. Fruits Not Seen Eaten by Treeshrews

Acorns: *Lithocarpus* spp. and *Castanopsis* spp	P	*S. hippurus; Callosciurus prevosti; Ratufa affinis; Sundasciurus brookei, S. lowii, Muntiacus muntjac*
	D	*T. tana?; Sundasciurus hippurus; S. lowii; Ratufa affinis; Presbytis rubicunda; Macaca nemestrina; Muntiacus muntiac; Tragulus* sp.
Canarium denticulatum	D	*Sundasciurus hippurus*; rats; *Calyptomena viridis; Anorrhinus galeritus; Ptilinopus jambu; Ducula aenea*; barbets; other broadbills
Diospyros spp.	D	*Tragulus* sp.; *Viverra tangalunga; Buceros rhinoceros*
Ficus, 10 December	D	*Muntiacus muntiac; Rhinoplax vigil*
Ficus (W2N225), 18–28 May	D	*Hylobates muelleri*; tree packed with birds, hornbills; *Argusianus*

Fruit	Site	Consumers
argus below *Nephelium rambutan-ake*, 13 October	D	*Pongo pygmaeus*
Mangifera sp.	D	*Pongo pygmaeus*
Microcos sp.	D	*Tragulus napu*
Pandanus sp.	D	*Sundasciurus lowii*
Ryparosa huleti	D	*Callosciurus prevosti; Lenothrix canus*
Willughbeia sp.	D	*Callosciurus prevosti; Sus barbatus; Lophura ignita*

APPENDIX VI

Response of Murid Rodents to the Masting Phenomenon of 1990–1991

During the monthly trapping for treeshrews at Danum Valley, traps were left open at night to evaluate murid rodent communities. All murids were marked by toe clipping and released, and recorded at recapture. From October to December, traps were left open for two nights each session, but because high murid capture numbers interfered with treeshrew captures, from early March (M1), traps were set for one night only. Elephants interfered with trapping in January, so the data for that month are not comparable. Four species of murid rodents dominated the captures: *Maxomys surifer, Maxomys rajah, Leopoldomys sabanus,* and *Niviventer cremoriventer.*

RESULTS AND DISCUSSION

The massive fruit masting that we recorded in September–November 1990 (fig. 3.5) was first followed by a small increase in rodent numbers (2.5–14% trapping success) and then, after four months, by an enormous increase (to 65% success), which was more or less maintained for the following seven months (fig. A-VI.1A). A species-by-species analysis of these results shows that every species showed a small early increase in captures that contributed to the October–November rise in captures, while the delayed rodent explosion that followed the masting was entirely due to a single species, *Maxomys rajah*. The peaks of fruit and rainfall in May 1991 were also followed a month later by a small increase in captures of all rodent species.

Because no *M. rajah* were captured in the first (5-day) trapping session in September 1990, I interpret the slow pattern of increase as reflecting a slow buildup of the population over several generations, from extremely low initial numbers. Whereas the other three species fluctuated around low but fairly constant num-

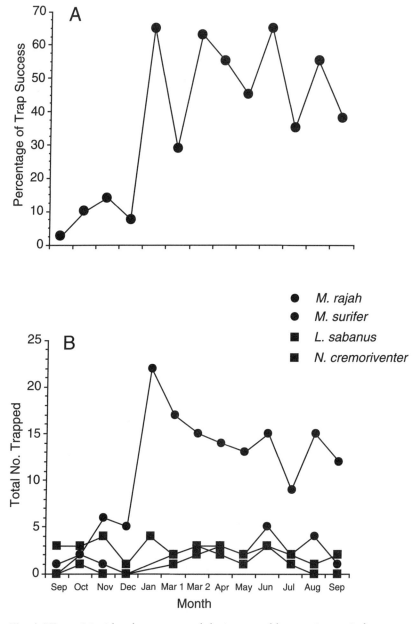

Fig. A-VI.1. Murid rodents captured during monthly trapping periods at Danum Valley. A, total murids trapped, expressed as percentage of trap success (animals captured/40 traps × nights of trapping); and B, total number of individuals caught, of the four most common species, not adjusted for differences in the number of nights of trapping (see text).

bers throughout the year, such that even the masting peak showed no outstanding population effect, *Maxomys rajah* continued at very high numbers until the end of the study. Of course, it is possible that none of the population variations among rodents was due to the fruit masting but instead to some other, unknown cause.

It is striking that the murid rodents showed a completely different annual population density pattern from treeshrews, without the clear tracking of fruiting or rainfall peaks. This implies a fundamentally different response to ecological factors by members of the two orders.

Bibliography

Allen, M. E. 1989. Nutritional aspects of insectivory. Ph.D. diss., Michigan State University.
Alterman, L., G. A. Doyle, and M. K. Izard, eds. 1995. *Creatures of the dark: The nocturnal prosimians.* New York: Plenum Press.
Anon. 1993. *Danum Valley Conservation Area: A checklist of vertebrates.* Kota Kinabalu: Innoprise Corporation.
Banks, E. 1931. A popular account of the mammals of Borneo. *J. Malayan Branch, Roy. Asiat. Soc.* 9: 1–139.
Bearder, S. K. 1987. Lorises, bushbabies, and tarsiers: Diverse societies in solitary foragers. In *Primate societies,* edited by B. B. Smuts, D. L. Cheney, R. M. Seyfarth, R. W. Wrangham, and T. T. Struhsaker, 11–24. Chicago: University of Chicago Press.
Benton, M. 1991. *The rise of the mammals.* New York: Crescent Books.
Bradbury, J. W., and S. L. Vehrencamp. 1976. Social organization and foraging in emballonurid bats. I. Field studies. *Behav. Ecol. Sociobiol.* 1: 337–81.
Broekhuizen, S., and F. Maaskamp. 1976. Behaviour and maternal relations of young European hares during the nursing period. In *Ecology and management of European hare populations,* edited by Z. Pielowski and Z. Pucek, 59–67. Warsaw: Polish Hunting Association.
Bronson, F. H. 1989. *Mammalian reproductive biology.* Chicago: University of Chicago Press.
Brosset, A. 1990. A long-term study of the rain forest birds in M'Passa (Gabon). In *Biogeography and ecology of forest bird communities,* edited by A. Keast, 259–75. The Hague: Academic Publishing.
Buchler, E. R. 1975. Food transit time in *Myotis lucifugus* (Chiroptera: Vespertilionidae). *J. Mammal.* 56: 252–55.

Burghouts, T., G. Ernting, G. Korthals, and T. De Vries. 1992. Litterfall, leaf litter decomposition and litter invertebrates in primary and selectively logged dipterocarp forest in Sabah, Malaysia. *Phil. Trans. Roy. Soc. Lond.* B 335: 407–16.
Butler, P. M. 1980. The tupaiid dentition. In *Comparative biology and evolutionary relationships of tree shrews,* edited by W. P. Luckett, 171–204. New York: Plenum Press.
Cadle, J. E., and J. L. Patton. 1988. Distribution patterns of some amphibians, reptiles, and mammals of the eastern Andean slopes of southern Peru. In *Workshop on Neotropical distribution patterns,* edited by P. E. Vanzolini and W. R. Heyer, 225–44. Rio de Janeiro: Academia Brasileira de Ciencias.
Cantor, T. 1846. Catalogue of Mammalia inhabiting the Malayan Peninsula and islands. *J. Asiat. Soc. Bengal* 15: 171–203, 241–79.
Carlsson, A. 1922. Über die Tupaiidae und ihre Beziehungen zu den Insectivora und den Prosimiae. *Acta Zool.* 3: 227–70.
Charles-Dominique, P. 1971. Eco-éthologie des Prosimiens du Gabon. *Biologia Gabonica* 7: 123–228.
———. 1975. Nocturnality and diurnality. In *Phylogeny of the primates,* edited by W. P. Luckett and F. S. Szalay, 68–88. New York: Plenum Press.
———. 1977. Urine-marking and territoriality in *Galago alleni* (Waterhouse, 1837)—Lorisoidea, Primates—A field study by radio-telemetry. *Z. Tierpsychol.* 43: 113–38.
Charles-Dominique, P., and S. K. Bearder. 1979. Field studies of lorisid behavior: Methodological aspects. In *The study of prosimian behavior,* edited by G. A. Doyle and R. D. Martin, 567–629. New York: Academic Press.
Chivers, D. J., and C. M. Hladik. 1980. Morphology of the gastrointestinal tract in primates: Comparisons with other mammals in relation to diet. *J. Morph.* 166: 337–86.
Churchfield, S. 1990. *The natural history of shrews.* Ithaca: Cornell University Press.
Clark, W. E. Le Gros 1925. On the skull of *Tupaia. Proc. Zool. Soc. Lond.* 1925: 559–67.
———. 1926. On the anatomy of the pen-tailed tree-shrew (*Ptilocercus lowii*). *Proc. Zool. Soc. Lond.* 1926: 1179–1309.
Clark, W. E. Le Gros. 1959. *The antecedents of man: an introduction to the evolution of primates.* Edinburgh: Edinburgh University Press.
Corbet, G. B., and J. E. Hill. 1991. *A world list of mammalian species.* Natural History Publications, London. Oxford: Oxford University Press.
———. 1992. *The mammals of the Indomalayan region: A systematic review.* Oxford: Oxford University Press.
Damuth, J. D. 1992. Taxon-free characterization of animal communities. In *Terrestrial ecosystems through time: Evolutionary paleoecology of terrestrial plants and animals,* edited by A. K. Behrensmeyer, J. D. Damuth, W. A. DiMichele, R. Potts, H. D. Sues, and S. L. Wing, 183–203. Chicago: University of Chicago Press.
Dans, A. T. L. 1993. Population estimate and behavior of Palawan tree shrew, *Tupaia palawanensis* (Scandentia, Tupaiidae). *Asia Life Sci.* 2: 201–14.

Davis, D. D. 1962. Mammals of the lowland rain-forest of North Borneo. *Bull. Nat. Mus. Singapore,* no. 31.
Dietz, J. M. 1984. Ecology and social organization of the maned wolf (*Chrysocyon brachyurus*). *Smithsonian Contrib. Zool.,* no. 392.
Doyle, G. A., and R. D. Martin, eds. 1979. *The study of prosimian behavior.* New York: Academic Press.
D'Souza, F. 1972. A preliminary field report on the lesser tree shrew *Tupaia minor.* In *Prosimian biology,* edited by R. D. Martin, G. A. Doyle, and A. C. Walker, 167–82. London: Duckworth.
D'Souza, F., and R. D. Martin. 1974. Maternal behaviour and the effects of stress in tree shrews. *Nature* 251: 309–11.
Dubost, G. 1988. Ecology and social life of the red acouchy, *Myoprocta exilis;* comparison with the orange-rumped agouti, *Dasyprocta leporina. J. Zool. Lond.* 214: 107–23.
Duplantier, J. M., P. Orsini, M. Thohari, J. Cassaing, and H. Croset. 1984. Echantillonage de populations de Muridés: Influence du protocole de piégeage sur l'estimation de paramètres démographiques. *Mammalia* 48: 129–41.
Eisenberg, J. F., and E. Gould. 1970. The tenrecs: A study in mammalian behavior and evolution. *Smithsonian Contrib. Zool.,* no. 27.
Elliot, O. 1971. Bibliography of the tree shrews 1780–1969. *Primates* 12: 323–414.
Emmons, L. H. 1975. Ecology and behavior of African rainforest squirrels. Ph.D. diss., Cornell University.
———. 1978. Sound communication among African rainforest squirrels. *Z. Tierpsychol.* 47: 1–49.
———. 1980. Ecology and resource partitioning among nine species of African rainforest squirrels. *Ecol. Monogr.* 50: 31–54.
———. 1982. Ecology of *Proechimys* (Rodentia, Echimyidae) in southeastern Peru. *Trop. Ecol.* 23: 280–90.
———. 1988. A field study of ocelots in Peru. *Rev. Ecol.* 42: 133–57.
———. 1991. Frugivory in treeshrews (*Tupaia*). *Amer. Nat.* 138: 642–49.
———. 1995. Mammals of rain forest canopies. In *Forest canopies,* edited by M. D. Lowman and N. M. Nadkarni, 199–233. San Diego: Academic Press.
———. N.d. Arboreal/scansorial insectivores. Unpublished ms.
Emmons, L. H., and A. Biun. 1991. Malaysian treeshrews. *Natl. Geogr. Res. Explor.* 7: 70–81.
Emmons, L. H., and F. Feer. 1997. *Neotropical rainforest mammals: A field guide.* 2d ed. Chicago: University of Chicago Press.
Emmons, L. H., J. Nais, and A. Biun. 1991. The fruit and consumers of *Rafflesia keithii* (Rafflesiaceae). *Biotropica* 23: 197–99.
Emry, R. J., and R. W. Thorington, Jr. 1982. Descriptive and comparative osteology of the oldest fossil squirrel *Protosciurus* (Rodentia: Sciuridae). *Smithsonian Contrib. Paleobiol.,* no. 47.
Fish, D. R., and F. C. Mendel. 1982. Mandibular movement patterns relative to food types in common tree shrews (*Tupaia glis*). *Amer. J. Phys. Anthropol.* 58: 255–69.
Flannery, T. F. 1990. *Mammals of New Guinea.* Carina, Australia: Robert Brown and Associates.

Flannery, T. F., and L. Seri. 1975. The mammals of southern West Sepik Province, Papua New Guinea: Their distribution, abundance, human use and zoogeography. *Rec. Austral. Mus.* 42: 173–208.

Fooden, J. 1964. Stomach contents and gastro-intestinal proportions in wild-shot Guianan monkeys. *Am. J. Phys. Anthropol.* 22: 227–32.

Foster, M. S. 1974. Rain, feeding behavior, and clutch size in tropical birds. *Auk* 91: 722–26.

Foster, R. B. 1982. Famine on Barro Colorado Island. In *The ecology of a tropical forest: Seasonal rhythms and long-term changes*, edited by E. G. Leigh, Jr., A. S. Rand, and D. M. Windsor, 201–25. Washington, D.C.: Smithsonian Institution Press.

Gautier-Hion, A., J.-M. Duplantier, R. Quris, F. Feer, C. Sourd, J.-P. Decoux, G. Dubost, L. H. Emmons, C. Erard, P. Hecketsweiler, A. Moungazi, C. Roussilhon, and J.-M. Thiollay. 1985. Fruit characters as a basis of fruit choice and seed dispersal by a tropical vertebrate community. *Oecologia* 65: 324–37.

Goldizen, A. W. 1987. Tamarins and marmosets: Communal care of offspring. In *Primate societies*, edited by B. B. Smuts, D. L. Cheney, R. M. Seyfarth, R. W. Wrangham, and T. T. Struhsaker, 34–43. Chicago: University of Chicago Press.

Goodman, S. M. 1994. The enigma of antipredator behavior in lemurs: Evidence of a large extinct eagle on Madagascar. *Int. J. Primatol.* 15: 129–34.

Gould, E. 1978. The behavior of the moonrat, *Echinosorex gymnurus* (Erinaceidae) and the pentail shrew *Ptilocerus lowi* (Tupaiidae) with comments on the behavior of other Insectivora. *Z. Tierpsychol.* 48:1–27.

Grant, T. R. 1983. *The platypus*. Kensington: New South Wales Press.

Greenberg, R., and J. Gradwohl. 1980. Leaf surface specializations of birds and arthropods in a Panamanian forest. *Oecologia* 46: 115–24.

Gregory, W. K. 1910. The orders of mammals. *Bull. Amer. Mus. Nat. Hist.* 27: 1–524.

Griffiths, M. 1978. *The biology of monotremes*. New York: Academic Press.

Guillotin, M. 1982. Rhythms d'activité et régimes alimentaires de *Proechimys cuvieri* et d' *Oryzomys capito velutinus* (Rodentia) en forêt Guyanaise. *Rev. Ecol.* 36: 337–71.

Han, K.-H., F. H. Sheldon, and R. B. Stuebing. N.d. Interspecific relationships and biogeography of the Southeast Asian treeshrews, with emphasis on Bornean species of *Tupaia*. Unpublished ms.

Happold, D. C. D., and M. Happold. 1978. The fruit bats of western Nigeria. *Nigerian Field* 43: 30–37.

Harrison, J. L. 1954. The natural food of some rats and other mammals. *Bull. Raffles Mus.* 25: 157–65.

———. 1955. Data on the reproduction of some Malayan mammals. *Proc. Zool. Soc. Lond.* 125: 445–60.

———. 1962. The distribution of feeding habits among animals in a tropical rain forest. *J. Anim. Ecol.* 31: 53–63.

Heaney, L. R., P. D. Heideman, E. A. Rickart, R. B. Utzurrum, and J. S. H. Klompen. 1989. Elevational zonation of mammals in the central Philippines. *J. Trop. Ecol.* 5: 259–80.

Hill, W. C. O. 1958. Pharynx, oesophagus, stomach, small and large intestine, form and position. *Primatologia: Handb. Primatenk.* 3(1): 139–207.

Holloway, J. D., A. H. Kirk-Spriggs, and Chey Vun Khen. 1992. The response of some rain forest insect groups to logging and conversion to plantation. *Phil. Trans. Roy. Soc. Lond. B* 335: 425–36.

Honacki, J. H., K. E. Kinman, and J. W. Koeppl. 1982. *Mammal species of the world: A taxonomic and geographic reference.* Lawrence, Kan.: Association of Systematics Collections.

Hose, C. 1893. *A descriptive account of the mammals of Borneo.* London: Edward Abbot.

Janzen, D. H. 1974. Tropical blackwater rivers, animals, and mast fruiting by the Dipterocarpaceae. *Biotropica* 6: 69–103.

Jenkins, F. A., Jr. 1974. Tree shrew locomotion and the origins of primate arborealism. In *Primate locomotion,* edited by F. A. Jenkins, Jr., 85–115. New York: Academic Press.

Johns, A. D. 1992. Vertebrate responses to selective logging: Implications for the design of logging systems. *Phil. Trans. Roy. Soc. Lond. B* 335: 437–42.

Kawamichi, T., and M. Kawamichi. 1979. Spatial organization and territory of tree shrews (*Tupaia glis*). *Anim. Behav.* 27: 381–93.

———. 1982. Social system and independence of offspring of treeshrews. *Primates* 23: 189–205.

Kenward, R. E. 1987. *Wildlife radio tagging: Equipment, field techniques, and data analysis.* London: Academic Press.

Kielan-Jaworowska, Z., T. M. Bown, and J. A. Lillegraven. 1970. Eutheria. In *Mesozoic mammals: The first two-thirds of mammalian history,* edited by J. A. Lillegraven, Z. Kielan-Jaworowska, and W. A. Clemens, 221–58. Berkeley: University of California Press.

Kleiman, D. 1977. Monogamy in mammals. *Quart. Rev. Biol.* 52: 39–69.

Klite, P. D. 1965. Intestinal bacterial flora and transit time of three Neotropical bat species. *J. Bacteriol.* 90: 375–79.

Klopfer, P. H., and K. J. Boskoff. 1979. Maternal behavior in prosimians. In *The study of prosimian behavior,* edited by G. A. Doyle and R. D. Martin, 123–56. New York: Academic Press.

Kobayashi, T., K. Maeda, and M. Harada. 1980. Studies on the small mammal fauna of Sabah, East Malaysia. I. Order Chiroptera and genus *Tupaia* (Primates). *Contrib. Biol. Lab. Kyoto Univ.* 26: 67–82.

Lambert, F. R. 1992. The consequences of selective logging for Bornean lowland forest birds. *Phil. Trans. Roy. Soc. Lond. B* 335: 443–57.

Lambert, F. R., and A. G. Marshall. 1991. Keystone characteristics of bird-dispersed *Ficus* in a Malaysian lowland rain forest. *J. Ecol.* 79: 793–809.

Lambert, F., and M. Woodcock. 1996. *Pittas, broadbills and asites.* Mountfield, U.K.: Pica Press.

Langham, N. P. E. 1982. The ecology of the common tree shrew, *Tupaia glis,* in peninsular Malaysia. *J. Zool. London* 197:323–44.

Lillegraven, J. A., Z. Kielan-Jaworowska, and W. A. Clemens, eds. 1979. *Mesozoic mammals: The first two-thirds of mammalian history.* Berkeley: University of California Press.

Lim, B. L. 1965. Food and weights of small animals from the First Division, Sarawak. *Sarawak Mus. J.* 12: 360–72.

———. 1967. Note on the food habits of *Ptilocercus lowii* Gray (Pentail treeshrew) and *Echinosorex gymnurus* (Raffles) (Moonrat) in Malaya with remarks on "ecological labelling" by parasite patterns. *J. Zool. Lond.* 152: 375–79.

Luckett, W. P., ed. 1980. *Comparative biology and evolutionary relationships of tree shrews.* New York: Plenum Press.

Lyon, M. W., Jr. 1913. Tree shrews: An account of the mammalian family Tupaiidae. *Proc. U.S. Natl. Mus.* 45: 1–188.

MacKinnon, J. R., and K. S. MacKinnon. 1980. The behaviour of wild spectral tarsiers. *Int. J. Primatol.* 1: 361–79.

McNab, B. K. 1963. Bioenergetics and the determination of home range size. *Amer. Nat.* 97: 133–40.

———. 1983. Ecological and behavioral consequences of adaptation to various food resources. In *Advances in the study of mammalian behavior,* edited by J. E. Eisenberg and D. G. Kleiman, 664–97. American Society of Mammalogists Spec. Publ. 7.

Marsh, C. W., and A. G. Greer. 1992. Forest land-use in Sabah, Malaysia: An introduction to Danum Valley. *Phil. Trans. Roy. Soc. Lond. B* 335: 331–39.

Marshall, A. G., and M. D. Swaine. 1992. Tropical rain forest: Disturbance and recovery. *Phil. Trans. Roy. Soc. Lond. B* 335: 323–462.

Martin, R. D. 1968. Reproduction and ontogeny in tree-shrews (*Tupaia belangeri*), with reference to their general behaviour and taxonomic relationships. *Z. Tierpsychol.* 25: 409–96, 505–32.

Mitchell, P. C. 1905. On the intestinal tract of mammals. *Trans. Zool. Soc. Lond.* 17: 437–537.

———. 1916. Further observations on the intestinal tract of mammals. *Proc. Zool. Soc. Lond.* 1916: 183–251.

Morton, E. 1975. Ecological sources of selection on avian sounds. *Amer. Nat.* 108: 17–34.

Muul, I., and B. L. Lim. 1971. New locality records for some mammals of West Malaysia. *J. Mammal.* 52: 430–37.

Newbery, D. McC., E. J. F. Campbell, Y. F. Lee, C. E. Ridsdale, and M. J. Still. 1992. Primary lowland dipterocarp forest at Danum Valley, Sabah, Malaysia: Structure, relative abundance and family composition. *Phil. Trans. Roy. Soc. London B* 335: 341–56.

Novacek, M. J. 1980. Cranioskeletal features in tupaiids and selected Eutheria as phylogenetic evidence. In *Comparative biology and evolutionary relationships of tree shrews,* edited by W. P. Luckett, 35–93. New York: Plenum Press.

Oftedal, O. T., D. J. Boness, and R. A. Tedman. 1987. The behavior, physiology, and anatomy of lactation in the Pinnipedia. In *Current mammalogy,* edited by H. H. Genoways, 175–245. New York: Plenum Press.

Overdorff, D. J., and M. A. Rasmussen, 1995. Determinants of nighttime activity in "diurnal" lemurid primates. In *Creatures of the dark: The nocturnal prosimians,* edited by L. Alterman, G. A. Doyle, and M. K. Izard, 61–74. New York: Plenum Press.

Payne, J. B. 1979. Synecology of Malayan tree squirrels. Ph.D. diss., Cambridge University.
Payne, J., C. M. Francis, and K. Phillipps. 1985. *A field guide to the mammals of Borneo*. Kuala Lumpur: Sabah Society.
Peterson, E. A., S. D. Wruble, and V. I. Ponzoli. 1968. Auditory responses in tree shrews and primates. *J. Aud. Res.* 8: 345–55.
Primack, R. B., and P. Hall. 1992. Biodiversity and forest change in Malaysian Borneo. *Bioscience* 42: 829–37.
Prince, J. H. 1956. *Comparative anatomy of the eye*. Springfield, Ill.: Charles C. Thomas.
Raemaekers, J. J., F. P. G. Aldrich-Blake, and J. B. Payne. 1980. The forest. In *Malayan forest primates*, edited by D. J. Chivers, 29–61. New York: Plenum Press.
Raffles, T. S. 1822. Descriptive catalogue of a zoological collection made on the account of the Honorable East India Company, in the island of Sumatra and its vicinity, under the direction of Sir Thomas Raffles, Lieutenant-Governor of Fort Marborough; with additional notices illustrative of the natural histories of those countries. *Trans. Linn. Soc. Lond.* 13: 239–74.
Rathbun, G. B. 1979. The social structure and ecology of elephant-shrews. *Adv. Ethol.* 20: 1–77.
Ray, J. C. 1996. Resource use patterns among mongooses and other carnivores in a Central African rainforest. Ph.D. diss., University of Florida.
Richardson, K. C., R. B. Stuebing, and H. K. Normah. 1987. Alimentary tract morphology and digesta transit of some Malaysian chiropterans. *Indo-Malayan Zool.* 4: 399–412.
Ridley, H. N. 1895. The mammals of the Malay Peninsula. *Nat. Sci.* 6 (35): 23–29.
Roberts, M. 1993. The minimalist motherhood of tree shrews. *Zoogoer* (July–August): 6–11.
Robinson, S. K., J. Terborgh, and C. A. Munn. 1990. Lowland tropical bird communities of a site in western Amazonia. In *Biogeography and ecology of forest bird communities*, edited by A. Keast, 229–58. The Hague: Academic Publishing.
Rodieck, R. W. 1986. The primate retina. In *Comparative primate biology*, edited by H. D. Steklis and J. Erwin, 203–78. *Neurosciences*, vol. 4.
Root, R. B. 1967. The niche exploitation pattern of the blue-gray gnatcatcher. *Ecol. Monogr.* 37: 317–50.
Ryan, J. N., G. K. Creighton, and L. H. Emmons. 1993. Activity patterns of two species of *Nesomys* (Muridae: Nesomyinae) in a Madagascar rain forest. *J. Trop. Ecol.* 9: 101–7.
Ryan, J. N., L. H. Emmons, E. Raholimavo, and G. K. Creighton. N.d. Non-primate mammals of Parc National de Ranomafana. Unpublished ms.
Samorajski, T., J. M. Ordy, and J. R. Keefe. 1966. Structural organization of the retina in the tree shrew (*Tupaia glis*). *J. Cell Biol.* 28: 489–504.
Sherman, P. T., and P. K. Eason. 1998. Size determinants in territories with inflexible boundaries: Manipulation experiments on white-winged trumpeters. *Ecology* 79: 1147–59.

Shriver, J. E., and C. R. Noback. 1967. Color vision in the tree shrew (*Tupaia glis*). *Folia Primatol.* 6: 161–69.

Simpson, G. G. 1945. The principles of classification and a classification of mammals. *Bull. Amer. Mus. Nat. Hist.* 85: 1–350.

Smith, A. T., and B. L. Ivins. 1983. Reproductive tactics of pikas: Why have two litters? *Canad. J. Zool.* 61: 1551–59.

———. 1984. Spatial relationships and social organization in adult pikas: A facultatively monogamous mammal. *Z. Tierpsychol.* 66: 289–308.

Smuts, B. B., D. L. Cheney, R. M. Seyfarth, R. W. Wrangham, and T. T. Struhsaker, eds. 1987. *Primate societies*. Chicago: University of Chicago Press.

Smythies, B. E. 1981. *The birds of Borneo*. 3d ed. Kuala Lumpur: Sabah Society and Malayan Nature Society.

Snyder, M., H. Killackey, and I. T. Diamond. 1969. Color vision in the tree shrew after removal of posterior neocortex. *J. Neurophys.* 32: 554–63.

Soini, P. 1988. The pygmy marmoset, genus *Cebuella*. In *Ecology and behavior of Neotropical primates*, edited by R. A. Mittermeier, A. B. Rylands, A. F. Coimbra-Filho, and G. A. B. de Fonseca, 79–129. Washington, D.C.: World Wildlife Fund.

Sprankel, H. 1961. Über Verhaltensweisen und Zucht von *Tupaia glis* (Diard 1820) in Gefangenschaft. *Z. Wiss. Zool.* 165: 186–220.

Sterling, E. J., and A. F. Richard. 1995. Social organization in the aye-aye (*Daubentonia madagascariensis*) and the perceived distinctiveness of nocturnal primates. In *Creatures of the dark: The nocturnal prosimians*, edited by L. Alterman, G. A. Doyle, and M. K. Izard, 439–51. New York: Plenum Press.

Stork, N. E. 1991. The composition of the arthropod fauna of Bornean lowland rain forest trees. *J. Trop. Ecol.* 7: 161–80.

Stuebing, R. B., and J. Gasis. 1989. A survey of small mammals within a Sabah tree plantation in Malaysia. *J. Trop. Ecol.* 5: 203–14.

Szalay, F. S., and G. Drawhorn. 1980. Evolution and diversification of the Archonta in an arboreal milieu. In *Comparative biology and relationships of tree shrews*, edited by W. P. Luckett, 133–69. New York: Plenum Press.

Tan Kong Beng. 1965. Stomach contents of some Borneo mammals. *Sarawak Mus. J.* 12: 373–85.

Tattersall, I. 1984. The tree-shrew, *Tupaia*: A "living model" of the ancestral primate? In *Living fossils*, edited by N. Eldredge and S. M. Stanley, 32–37. New York: Springer Verlag.

Terborgh, J. 1983. *Five New World primates*. Princeton: Princeton University Press.

Thomas, O. 1910. Two new mammals from the Malay Peninsula. *Ann. Mag. Nat. Hist.*, ser. 2, 5: 424–26.

Thomas, O., and R. C. Wroughton. 1909. On mammals from the Rhio Archipelago and Malay Peninsula collected by Messrs. H. C. Robinson, C. Boden Kloss, and E. Seimund, and presented to the National Museum by the Government of the Federated Malay States. *J. Fed. Malay States Mus.* 4: 99–129.

Wade, P. 1958. Breeding season among mammals in the lowland rain-forest of North Borneo. *J. Mammal.* 39: 429–33.

Wharton, C. H. 1950. Notes on the Philippine tree shrew *Urogale everetti* Thomas. *J. Mammal.* 31: 352–54.
Whittow, G. C., and E. Gould. 1976. Body temperature and oxygen consumption of the pentail tree shrew (*Ptilocercus lowii*). *J. Mammal.* 57: 754–56.
Wilson, D. E. 1993. Order Scandentia. In *Mammal species of the world: A taxonomic and geographic reference,* edited by D. E. Wilson and D. M. Reeder, 131–33. Washington, D.C.: Smithsonian Institution Press.
Wright, P. C. 1985. The costs and benefits of nocturnality for *Aotus trivirgatus* (the night monkey). Ph.D. diss., City University of New York.
Wright, P. C., and L. B. Martin. 1995. Predation, pollination and torpor in two nocturnal prosimians: *Cheirogaleus major* and *Microcebus rufus* in the rain forest of Madagascar. In *Creatures of the dark: The nocturnal prosimians,* edited by L. Alterman, G. A. Doyle, and M. K. Izard, 45–60. New York: Plenum Press.

Index

Page numbers in **bold** refer to figures; page numbers in *italic* refer to tables.

Absentee system of parental care. *See* Maternal care
Activity patterns, 119–21; active period duration, 110–13, *111*, *112*, 114, 212; active period duration, travel distances and, 144; of common shrews, 122; evolutionary aspects, 121–22; measuring, 122–23; morphology and, 121; rain effects on, 116–19; rest periods, *112*, 114–15, 115; resting behaviors, 113–15; time of day, 110, 113
Alarm calls, 175, 202, 203–4; social organization and, 208; tail flicking with, 203, 204; types of, 203. *See also* Predation
Alarm displays: tail movements, 15. *See also* Vocalization
Arboreal leaping, 45–46
Arboreality: evolutionary aspects, 223; home range and, 212–13; insectivory and, 218, 220, 223; in mammals, 220; morphological adaptations for, 8, 46, 223, 224; terrestriality dichotomy, 223–24; of treeshrews, 45–46, 220, 223
Archonta, 4
Arthropods: abundance, treeshrew breeding and, 193; bird rivalry for, 85–86, 86; foraging for, 193–94; as prey, 75, 76–78, *77*, **78**

Bats: Emballonurid bat *(Rhynchonycteris naso)*, social organization, 163–64; frugivorous, digestive tracts, 13; frugivorous, fiber-spitting behavior, 60; frugivorous, fruit processing, 59, 60–61; group huddling, 108; insectivorous, digestive tracts, 13; insectivorous, feeding habits, 218
Biodiversity: islands and, 17
Biomass: estimation method, 141; population density and, 140–42
Birds: babbler *(Timalidae)*, foraging behaviors, 85, 86; digestive tracts, 12–14; feeding similarities to treeshrews, 79; foraging association with squirrels, 89; foraging association with treeshrews, 86–89, **88**, **89**, 166; frugivorous, 213; monogamy in, 163; pitta *(Pitta baudi, P. venusta)*, alarm call similarity with treeshrews, 204; pitta *(Pitta baudi, P. venusta)*, as rivals for arthropods, 85; rain effects on breeding, 121; raptors, as predators, 206–7; as rivals for fruit, 82–84; as rivals for invertebrates, 85–86, 86
Birth: litter size, *176*, 176–77; seasonality, 177–81, *178*, **180**
Biun, Alim, 26, 27, 95, 100
Body mass: home range size and, 125–28, **129**, 211–12; seasonal patterns,

261

Body mass (*continued*)
185–87, 189; treeshrew species comparisons, 21
Body measurements: male vs. female, 18–21, 20, 187, 215; treeshrew species comparisons, 21
Borneo: biological diversity in, 24; food supply, population sizes and, 142, 144; Sabah, 24; Sabah Softwoods (tree plantation), 50–51; study sites (map), 25; treeshrew diversity in, 4, 17, 24, 225. *See also* Danum Valley; Poring; Tropical forests
Breeding. *See* Reproduction
Bronson, F. H., 169
Burrows: safety of, 207–8; of *T. longipes*, 94, 95, 95–96; termite prey and, 107
Butler, P. M., 10

Carlsson, A., 3
Clark, W. E. Le Gros, 3, 15
Common shrew *(Sorex araneus)*: activity patterns, 121–22; obesity in captivity, 121. *See also* Elephant shrew; Shrew
Competition for fruit: among treeshrews, 81–82, 214; with other animals, 82–84, 214
Crusafontia: treeshrew resemblance to, 8, 9

Dans, A. T. L., 6, 142
Danum Valley: fruit species collected at, 232–36; fruiting phenology, 36, 62–64, 63, 64; logged forest, species composition, 50–51; logged forest, treeshrews in, 34, 48–50, 50; rainfall, 34–36, 35, 116, 117, 118, 120–21; site characteristics, 31, 34, 34; trails, 32
Davis, D. D., 44–45, 179
Dendrogale melanura (Smooth-tailed treeshrew), 47
Dentition. *See* Teeth
Diet, 53, 78–80; ecological consequences of, 80–81; ecological separation of species and, 210–11, 211, 220; squirrels' compared, 82–84. *See also* Foraging; Frugivory; Insectivory
Digestive tracts, 12–14, 59, 220; compared, 13
Dipterocarp trees, 25, 26, 33
Distances traveled. *See* Home ranges; Travel distances
D'Souza, F., 45, 120, 170
Dzuhari, Mohammad, 31

Elephant shrew: absentee maternal care, 197–98; activity patterns, 122, 221; digestive tract, 12; diurnality, 219; foraging, 168; monogamy, 162, 167; ranging behaviors, 142, 143; sensory capacities, 17; social systems, similarities to treeshrews, 221; territoriality, 167, 168
Elevational distribution, 46–48, 47
Ellis, William, 1
Emballonurid bat *(Rhynchonycteris naso)*: social organization, 163–64
Evolution: of nesting by mammals, 107–8; primitive aspects of treeshrews, 4–5, 9, 23, 223, 224

Feces: collection methodology, 69, 89–90; containing fruit, 65
Feeding behavior: fiber spitting, 59–60, 60; for fruits, 56, 59–61, 60, 80; mastication, 60; salivation, 60; temporal patterns, 61–62, 62. *See also* Foraging
Field methods. *See* Methodology
Food habits. *See* Diet
Food supply: population sizes and, 142, 144; territory size and, 193, 215
Foraging, 212; arboreal, 39, 44, 212–13; arboreal, on tree trunks, 66, 218–19; bird associations with treeshrews, 86–89, 88, 89; ecological separation of similar species and, 210–11, 220–21; facilitated by Prevost's squirrel *(Callosciurus prevosti)*, 83; for fruit, 59–61, 193; fruit trees as specific destinations, 136, 137–38, 166; fruit trees' influence on movements, 53–56, 54, 166, 193; heights for, 44–45, 212; home range size and, 212, 212–13; for invertebrates, 66–69, 80, 193–94, 215; needs, habitat differences and, 215–16; in pairs, predation and, 208–9; on stream banks, 138–40, 139; surfaces, 212; territory size and, 215. *See also* Feeding behavior
Frugivore: digestive tracts, 13. *See also* Frugivory
Frugivory, 79, 213; competition for fruit, among treeshrews, 81–82, 214; competition for fruit, with other animals, 82–84, 214; feeding behavior, 56, 59–61, 60, 80; fiber-spitting behavior, 59–60, 60; foraging strategy, 59, 59; fruit sizes eaten, 56, 57; fruit species eaten, 53–57, 55, 64, 64–66, 65, 137–38, 244–46; generic differences in, 58–59; home range size and, 213;

morphology and, 61; seasonal abundance of fruit and, 62–64, **63**, **64**, 79; social organization and, 214–15; species differences in, *71*

Fruit: fig (*Ficus* spp.), 56, 64–65, 194, 213; in insectivore diets, 218; species at Danum Valley, 232–36; species eaten, 53–57, **55**, 64, 64–66, **65**, 137–38, 244–46; transit times, *59*, 59, 121

Fruit trees: foraging movements and, 53–56, **54**; as specific destinations, 136, 137–38, 166; time spent at, travel distances and, *137–38*

Fruiting phenology: body mass seasonal patterns and, 189; Danum Valley, 36, 62–66, **63**, **64**; eating patterns and, 79; masting event of September 1990, murid rodents' response to, 194, 247–49; masting event of September 1990, treeshrews' response to, 184, 194; reproduction patterns and, 183, 183–84, 193; sampling methods, 36, 64, 184

Galago (*Galagoides* spp.): maternal care of young, 108; nesting behavior, 108; territoriality, 162

Gasis, Elaine Campbell, 31, 49, 50–51, 232

Gould, E., 164

Guillotin, M., 129

Habitat, 38; elevational limits, 46–48, *47*; foraging needs and, 215–16; height distribution, 39; ranges, 46; substrate use, 43–44, *44*. *See also* Tropical forests

Harrison, J. L., 179

Hearing, 16–17; primates and treeshrews compared, *16*

Home ranges: arboreality and, 212–13; body mass and, 125–28, **129**, 211–12; defined, 125; displacements from, 192; of elephant shrews, 142, *143*; frugivory and, 213; gender differences in, **129**, *129*, *133*, 133, *134*; of primates, 142–44, *143*; size, 125–29, *126*, *127*, **128**, 211–13, *212*; size, as biogenetic indicator, 131–35; size, foraging and, 212–13; social organization and, 147, **148**, **149**; of squirrels, 142, *143*; travel distances and, *130*, 131–35, *132*, *133*, *134*, 135, 212; travel patterns, **133**. *See also* Travel speeds

Hussain, Hajinin, 26, 27

Insectivora: digestive tracts, 12; teeth, 11–12. *See also* Mammals

Insectivory, 79, 214; arboreality and, 218, 220, 223; birds' feeding similarities, 79; foraging behavior, 66, 80; fruit-eating and, 218; invertebrate prey species, 69–70, *71*, *72*, 75–76, 240–43; seasonal abundance of invertebrates, 76–78, *77*, *78*; social organization and, 215; species differences, *71*, *72*, 79; terrestriality and, 220

Insects. *See* Arthropods

Invertebrates: competition with other animals for, 84–86; prey species, 69–70, *71*, *72*, 75–76, 240–43; seasonal abundance, 76–78, *77*, *78*; seasonal abundance, reproduction patterns and, 183, 184

Islands: biodiversity roles, 17

Johns, A. D., 51

Juveniles: emergence from the nest, 173, 195; growth rates, 184–88, *185–87*; nurslings, 171–72, *172*; odor of, 198–99, 208

Kawamichi, M. and T., 45, 142, 161–62, 164, 165, 179, 215

Langham, N. P. E., 142, 182, 188–89, 192, 193, 194

Large treeshrew. *See Tupaia tana*

Lesser treeshrew. *See Tupaia minor*

Life history, 193–95; methodology, 200; patterns of terrestrial treeshrews, 192–93. *See also* Maturation

Litter size, 176, *176–77*

Lizard: as prey, 70, *71*, 76

Luckett, W. P., 3

Lyon, M. W., Jr., 3, 146

Mammals: absentee maternal care in, 197–98; activity cycles, 122; arboreality in, 220; arboreality-terrestriality dichotomy in, 223–24; diurnal, 219; ecological separation of similar species, 210–11, 220–21; home range size as bioenergetic indicator, 131; insectivory, in arboreal mammals, 218; life span, reproduction and, 169; monogamy in, 162, 163; nesting, evolutionary aspects, 107–8; nocturnal, 208, 219; as predators, 205–6, 206; primitive, treeshrews and, 4–5, 9, 23, 223, 224; ranging behaviors, 142–44, *143*; as rivals for fruit, 82–

Mammals (*continued*)
84; sensory capacities compared to treeshrews, 17; solitary territorial, social organization, 162–63; species found on study plots, 237–39. *See also* Insectivora

Marsupials: activity cycles, 122; foraging, 218; insectivorous, digestive tracts, 12; sensory capacities, 17

Martin, D., 5, 6, 162, 169, 179

Mastication, 60

Maternal care: absentee system, 6, 175, 225; absentee system, as evolutionary holdover, 196, 199; absentee system, features of, 169, 195; absentee system, functional reasons for (speculated), 196–99, 197; absentee system, in mammals, 197–98; absentee system, predation and, 198–99, 204, 208; in captivity, 170, 195; emergence of young from nest, 173, 175–76, 195; functional reasons for (speculations), 107, 196–99; methodology for study, 199–200; weaning, 170, 196

Maturation: age at first reproduction, 188–89; growth rates, 184–88, 185–87. *See also* Life history

Methodology, 36–37; activity pattern measurement, 122–23; biomass estimations, 141; capture effects, 200–201, 229–30; feces collection, 69, 89–90; fruit sampling, 36, 64, 184; handling treeshrews, 229; home range estimation, 124–25; life history data, 200; maternal care observations, 199–200; nest locating, 108–9, 199; observation, 38, 51–52; population density estimations, 140–41; problems encountered, 229–30; radio-tracking, 108–9, 110, 122–23, 230–31; radio-tracking, maternal care study, 199–200; radio-tracking, travel distance estimates from, 129–30; trapping, 227–29, 228

Mitchell, P. C., 12–14

Mongoose: as rival for invertebrate prey, 84

Monkey: hearing compared to treeshrews, 16. *See also* Primates

Monogamy, 163, 215; facultative, 162; obligate, 163; in *Ptilocercus*, 163; in tupaids, 161–62, 215–16, 222

Morphology, 1–2, 2, 8–10; arboreal adaptations, 8, 46, 223, 224; digestive tract, arboreality and, 224; digestive tracts, 12–14, 13, 59, 121;

ear pinnae, 16; foraging adaptations, 79; fruit processing and, 61; primitivity of, 222

Mountain treeshrew. *See Tupaia montana*

Nesting: by mammals, evolution of, 107–8

Nesting sites, 91–101, 104; burrows, termite prey and, 107; locating methods, 108–9. *See also* Sleeping sites

Nests: building, 103; safety of, 107, 207–8; structure, 104, 105–7; time of return to, 111, 113, 119, 120, 121

Noise: response to, 2, 16

Nurslings. *See* Juveniles

Olfaction, 2, 15–16; in foraging for invertebrates, 66; scent marking of territories, 164

Palawan treeshrews: population densities, 142

Parental care. *See* Maternal care

Pentail treeshrew. *See Ptilocercus lowii*

Philippine treeshrew (*Urogale everetti*): nesting sites, 91–92

Philippines: treeshrews in, 4, 6

Plain treeshrew. *See Tupaia longipes*

Population densities, 141, 141–42; biomass and, 140–42; estimation methods, 140–41

Population sizes: food supply and, 142, 144

Poring: site characteristics, 27–28, 29, 30; trails, 15; trap sites, 15

Predation, 204–5; defenses against, 202–3; foraging in pairs and, 208–9. *See also* Alarm calls

Predators, 205–7, *206*

Primates: dropping in response to danger, 207; monogamy in, 163; nesting behaviors, 108; ranging behaviors, 142–44, *143*; social organization, diurnal species, 208; social organization, nocturnal species, 208; treeshrew ancestry and, 4

Ptilocercinae (subfamily), 7, 218–19, 225

Ptilocercus lowii (Pentail treeshrew), 7, 18, 21–22, 218; active period duration, 110–11, 111, 112, 119, 120, 121; activity patterns, rainfall and, 116, 119, 120; alarm calls, 175, 204, 208; alarm displays, 15; arboreal pathways, 140, 218; arboreality, 45–46; biomass, 141; body mass, 21; body measurements, 20; competition for invertebrates, *86*, 86; danger, response to potential, 202; foraging, 39, 40, *212*, *213*, 218–19; foraging, for invertebrates, 66, *80*; foraging, movement patterns, 44, *44*; foraging, on tree trunks, 66, *80*, *212*, 218–19; foraging, solitary, 146, *163*; frugivory, 71, 79, 218; group huddling, 164; habitat height distribution, 39, 40; habitat substrate use, 42–43, *44*; habitat tolerance, 217; hearing, 16–17; home range, size, *126*, *127*, 127, **128**, 128, **132**, *212*, 212; home range, social organization and, 147, **148**; home range, travel distances and, *132*, *133*, *135*; insectivory, 66, *80*, 218–19; invertebrate prey, 70, 71, 73, *80*; litter size, 177; locomotion, 8–10, 218; maternal care, 175–76, 196; maternal care, alarm calls and, 204; maternal care, nest-sharing after juveniles' emergence, 175–76, 196; monogamy, 163; morphology, arboreal adaptations, 8–9, 45; nest safety, 107, 207; nest structure, *104*; nesting behavior, 104–5, 175–76, 196; nesting sites, 91, 92, 93, *104*, 104–5; nocturnal activity, 110; noise, response to, 17; population density, *141*, 141; primitive morphology, 222; reproductive histories, 177, *178*; rest periods, *112*, **115**; resting behavior, 114; sleeping sites, *102*, 105; sleeping sites, in captivity, 164; social organization, 145–47, 163–64, 221–22; social organization, alarm calls and, 208; social organization, home range and, 147, **148**; teeth, 11, **12**; travel distances, *130*; travel distances, home range size and, *132*, *133*, *135*; travel speeds, *130*; vibrissae, 17, 218; vision, 15, 218

Radio-tracking, 108–9, 110, 122–23, 199–200, 230–31; travel distance estimates from, 129–30
Rainfall: activity patterns and, 116–19; bird breeding and, 121; body mass seasonal patterns and, 189; Danum Valley, 34–36, *35*, 116, *117*, *118*, 120–21, 183; reproduction patterns and, 183, *183*, 184, 194
Ranging behaviors: ecological separation of treeshrew species, 210–11, **211**; mammals, 142–44, *143*
Raptors: as predators, 206–7
Reproduction: arthropod abundance and, 193; birth seasonality, 177–81, *178*, 180; breeding patterns, 177–78, 192; in captivity, 177–79, 193, 214; environmental correlates, 183–84, 193; food supply and, 193–95, 214; fruitfall and, 183, 183–84, 193, 214; litter size, *176*, 176–77; output of individual females, 181–82, *182*; rainfall and, 183, *183*, 184, 194; seasonal activity, 177, *178*, **179**, 181; sexual maturity, ages at, 188–89; sexual maturity, in captivity, 188; sexual maturity, food supply and, 193
Reptiles: as predators, 205, 206
Rest period durations, 114–15, **115**
Resting behaviors, 113–15
Resting locations, 113
Roberts, Miles, 169–70
Rodents: murid, captures, 247, *248*; murid, fruit masting and, 247–48; murid, population density, 249

Saimiri sciureus (squirrel monkey): digestive tract, *13*; home range, 144
Salivation, 60
Scandentia, 3, 4, 217; arboreality, 45–46
Scats. *See* Feces
Scent marking, 16, 149, 161, 162, 164
Seals: absentee maternal care, 197–98
Sensory capacities, 14–17
Sheldon, F., 88
Shrew ("true shrews"): activity patterns, 121–22; foraging strategies, 85; territoriality, 167–68. *See also* Common shrew; Elephant shrew
Sleeping sites, 106; examination of, 108; use of, 101–2, *102*. *See also* Nesting sites
Slender treeshrew. *See Tupaia gracilis*
Smell. *See* Olfaction
Smooth-tailed treeshrew *(Dendrogale melanura)*, 47
Social organization: alarm calls and, 208; defined, 145; home range and, 147, **148**, **149**; monogamy, 161–63, 222; of treeshrew species, 161–64; of true shrews, 167–68
Squirrel: alarm calls, 204; arboreality in, 220; diet compared to treeshrews, 82–84; foraging association with birds, 89; foraging on tree trunks, 219; fruit foraging of treeshrews facilitated by, 83; movement noisiness, 203; nesting burrows, termite prey and, 107; ranging behaviors, 142, *143*; reproduction seasonality, 181; as rival for fruit, 81, 82–84,

Squirrel *(continued)*
220; as rival for invertebrates, 84–85; species in *T. montana* elevational ranges, 48; in treeshrew home range, 48; treeshrew similarities, 1, 219
Squirrel monkey *(Saimiri sciureus)*: digestive tract, 13; home range, 144
Steiner, Warren, 69
Stream banks: as travel routes, 138–40, *139*
Study sites (map), 25. *See also* Borneo
Stuebing, Robert B., 49, 50–51

Teeth, 10–12, *11*, 220; of insectivorous mammals, 11–12; tribosphenic, 10, *12*
Termites: as prey, nesting burrows and, 107; as *T. longipes* prey, 74, 107, 221
Terrestriality: arboreality dichotomy, 223–24; insectivory and, 220
Territorial behaviors, 161, 166
Territoriality, 47–48, 165–66; defined, 145; ecological aspects, 166; foraging needs and, 215; of true shrews, 167–68; of *Tupaia* species, 165–66
Territories: breakdown of, 166; displacements from, 167, 192; formation of, 164; overlapping, 47–48; size, 195, 215; size, food supply and, 193, 215; solitary, 162; spatial arrangement, 166; of *Tupaia* species, 147–61, *149, 152, 154, 155, 157, 158–59*. *See also* Home ranges
Trapping, 227–29, *228*; in logged and unlogged forests, 49, 49–50; Poring sites, 15; of *T. minor*, 41, 50
Travel distances (daily), 129–30, *130, 133, 134*; active period durations and, 144; gender differences, *130*, 130, *133, 134*; home range size and, 130, 131–35, *132, 133, 134, 135*, 212; rate of travel and, *130*, 130; time spent at fruit trees and, *137–38*
Travel patterns, 133, 136
Travel routes: along stream banks, 138–40, *139*; arboreal pathways, 140
Travel speeds, *130*, 130; distances traveled and, 129–30, *130*; gender differences, *130*, 130. *See also* Home ranges
Tree plantations: *tupaia* in, 50–51
Treeshrew: arboreal-terrestrial dichotomy invalid for, 223–24; ecological separation between species, 210–11, *211*, 220; primitive mammals and, 4–5, 9, 23, 223, 224; species similarities and differences, 210–11; species studied, 17, **18–19**, 20; taxonomic classification, 1, 2–4, 7, 23, 199, 217. *See also Ptilocercus*; *Tupaia*
Tropical forests: cover density, treeshrews' preference for, 140; destruction, threat to treeshrews, 216–17; dipterocarp trees, 25, 26, 33; euphorb trees, 33; logged areas, species composition, 50–51; logged areas, treeshrews in, 34, 48–50, *50*, 215; population differences and, 51; in Sabah, Borneo, 25–29, 26, 29, 30, 33–34, *34*; tree plantations, 50–51, 215
Tupaia: active period duration, 110–13, *111*, *112*, 114, 119, 120; active period duration, gender differences, *112*, 113; activity patterns, rainfall and, 115, 117–19; alarm displays, 15; arboreal leaping, 45; arboreal species, characteristics, 8–10, 223; climbing adaptations, 8; competition among species for fruit, 81–82, 214; diurnal activity, 110, 113, 119; ecological characteristics, 8–10; fossils, 7–8; habitat differences, foraging needs and, 215–16; habitat elevational ranges, 46–48, *47*; hearing, 16; life history patterns, terrestrial species, 192–93; litter size, *176*, 176; locomotion, 2, 10; maternal care, nursing visits, 175; maternal care, weaning, in captivity, 170; monogamy, 163; nest safety, 107; nesting behavior, 105; pair formations, 164; parental care, 6; populations in different forest types, 51; sleeping sites, 106; social organization, 161–63, 222; social organization, monogamy, 161–62, 163, 222; studies of, 5–6; survivorship of individuals studied, 189–93, *191*; teeth, 10–12, *11*, 220; territorial overlap among species, 47–48; territoriality, 165–66; travel, cover density and, 140; in tree plantations, 50–51, 215; visual behavior, 15. *See also specific topics*
Tupaia dorsalis, 51; terrestrial habits, 44
Tupaia glis, 4, 5; foraging heights, 44, 45; geographic distribution, 217; growth rates, 188–89; hearing, 16; home range, 142, 166, 167; litter size, 177; mastication, 60; maternal care, absentee, 169; nest structure, 105, 106; nesting sites, 91; population

density, 195; reproduction, breeding age, 189; reproduction, in captivity, 179; reproduction, output of individual females, 182; reproduction, seasonality of, 179–81, 193, 194; social organization, 161–62; studies of, 5–6; survivorship of individuals, 192; territory size, 195, 215; vision, 14

Tupaia glis belangeri, 5
Tupaia glis chinensis, 5, 177
Tupaia glis ferruginea, 177
Tupaia gracilis (Slender treeshrew), 19, 22; active period duration, 111, 112; alarm calls, 203; bait fruit, competitive behavior for, 81–82; biomass, 141; birth seasonality, 180; body mass, 21; body mass, home range size and, **129**, 211; body mass, seasonal patterns, **185**, 189; body measurements, 20; competition for fruit, 81, 214; competition for invertebrates, 85, 86; digestive tract, 12, *13*; feeding behavior, temporal patterns, 61–62, *62*; foraging, heights for, 41, 45; foraging, for invertebrates, 67, *80*; foraging, solitary, 152–53; fruit species eaten, *137*, 213; fruit trees, influence on movements, 54; fruit trees, as specific destinations, 136; growth rates, **186**; habitat, elevational limits, 47, *47*; habitat, height distribution, 39, 41–42; habitat, substrate use, 43, 44; home range, body mass and, **129**, 211; home range, gender differences, **129**, *129*; home range, size, *126*, *127*, **127**, *128*, **128**, *132*, *133*, *212*, **212**; home range, social organization and, 150–52, *152*; home range, travel distances and, 132, 133, 135; home range, travel patterns, *133*; invertebrate prey, 70, 71, 72, 73, 76; litter size, **176**; in logged vs. unlogged forest, 49; nest safety, 207; nest structure, *104*; nesting sites, 93–94, *104*; population density, 141; reproduction, output of individual females, 181–82, **182**; reproductive activity, 177, **178**, **179**; rest periods, 112, **115**; rest periods, gender differences, 112, **114**; sleeping sites, 101, *102*; social organization, 163, 150–53; social organization, groupings, *150*; survivorship of individuals studied, 190, **191**; territoriality, 151–53, *152*, 163; territory formation, 163; travel distances, *130*; travel distances, home range size and, 132, 133, 135; travel distances, time spent at fruit trees and, 136, *137*; travel speeds, *130*, **130**; tree plantations, absence in, 50, 215

Tupaia longipes (Plain treeshrew), 4, 18, 22; active period duration, 111, *112*, *113*, *119*; alarm calls, 203, *204*; bait fruit, competitive behavior for, 81–82, *83*; biomass, *141*; birth seasonality, 180; body mass, 21; body mass, female vs. male, **186**, 189; body mass, home range size and, **129**, 211; body mass, seasonal patterns, **186**, 189; body measurements, 20; burrows, 94, 95, *95*–*96*; capture effects, 200; competition for fruit, 81; competition for invertebrates, 85, 86; feeding behavior, temporal patterns, 62; fiber-spitting behavior, 60, *60*; foraging, heights for, 44, *45*; foraging, for invertebrates, 68, *80*; fruit species eaten, *138*, 213; fruit trees, influence on movements, 54; fruit trees, as specific destinations, 136; geographic distribution, 217; growth rates, **184**, **186**, 188; habitat elevational limits, 47, *47*; habitat height distribution, 39, 42; habitat substrate use, 43, 44; home range, body mass and, **129**, 211; home range, gender differences, **129**; home range, size, *126*, *127*, **128**, *128*, **132**, *212*; home range, social organization and, 153–56, *154*, *155*; home range, travel distances and, 132, 133, 135; invertebrate prey, 70, 71, 72, 74, 78; litter size, **176**; in logged vs. unlogged forest, 49; longevity, 207–8; nest safety, 207–8; nest structure, 94, *96*, *104*, 105; nest structure, termite prey and, 107; nesting sites, 94–97, *104*, **104**; population density, *141*; population size, 142; reproduction, age at first, 188; reproduction, output of individual females, 181–82, **182**, 194; reproductive activity, 177, **178**, **179**; rest periods, *112*, **114**, **115**; sleeping sites, *102*, **102**; social organization, 153–56, 162, 166–67; social organization, groupings, *150*; social organization, home range and, 153–56, *154*, *155*; as subspecies of *T. glis*, 22; survivorship of individuals studied, 190, **191**, 192; territorial overlap with other treeshrews, 48; territoriality, 153–55, *154*, *155*; territoriality,

Tupaia longpipes (continued)
gender and, 155–56; territory formation, 163, 164–65; travel distances, *130*; travel distances, home range size and, **132, 133, 135**; travel distances, time spent at fruit trees and, 136, *138*; travel speeds, *130*, 130; in tree plantations, 51, 215

Tupaia minor (Lesser treeshrew), 19, 22; active period duration, 111, *112*, 120; alarm calls, 204; arboreal pathways, 140; biomass, *141*; body mass, 21; body mass, home range size and, *129*, 211; body measurements, *20*; breeding patterns, 177; competition for fruit, 81, 214; competition for invertebrates, 84–85, *86*; danger, response to potential, 202; digestive tract, *13*; feeding behavior, temporal patterns, 61; foraging, 39, 212, 213, 220; foraging, association with birds, 86–89, **88, 89**; foraging, for invertebrates, 66–67, *80*, 84; foraging, movement patterns, 43–44, *44*; foraging, in pairs, *150*, 208; foraging, social organization and, 147–49; frugivory, 220; fruit bait, competitive behavior for, 81–82; fruit species eaten, *137*, 213; fruit transit time, *59*, 59; fruit trees, as specific destinations, 136; geographic distribution, 217; growth rates, **185, 188**; habitat elevational limits, *47*; habitat height distribution, 39, 40–41, 44; habitat substrate use, 43, 44; home range, body mass and, *129*, 211; home range, size, *126*, *127*, **127, 128, 132**, 212, 212; home range, social organization and, 147, **149**; home range, travel distances and, **132, 133, 135**, *135*; home range, travel patterns, 133; invertebrate prey, 70, 71, 72, 73; litter size, *176*; in logged vs. unlogged forest, 49, 50, 51, 216; maternal care, 170; maternal care, absentee, 169; movement, silence of, 203; nest safety, 207; nest structure, *104*; nesting sites, 92–93, 103, *104*; population density, *141*, 141; predation, strategies against, 204, 207; reproduction, output of individual females, **182**; reproductive activity, 179, *179*; rest periods, *112*, 115; resting locations, 113–14; skull, 11; sleeping sites, *102*; social organization, 147–50, 162; social organization, groupings, *150*; social organization, home range and, 147, **149**; survivorship of individuals studied, 190; territorial behaviors, 161; territorial overlap with other treeshrews, 48; territories, 149, 150; trapping of, 41; travel along stream banks, 140; travel distances, *130*; travel distances, home range size and, **132, 133, 135**, *135*; travel distances, time spent at fruit trees and, 136, *137*; travel speeds, *130*; in tree plantations, 50, 215; vision, 14

Tupaia montana (Mountain treeshrew; Montane treeshrew), 23; active period duration, 111, *112*, 119; alarm calls, 203, 204; biomass, *141*; body mass, 21; body mass, home range size and, *129*; body measurements, *20*; competition for invertebrates, 85; digestive tract, *13*; foraging, heights for, 44; foraging, for invertebrates, 67–68, 79, *80*; foraging, in pairs, 208; fruit species eaten, *65*; habitat, 28–29, *30*; habitat elevational limits, 47, 48; habitat height distribution, 39, 42; habitat substrate use, 43, 44; home range, body mass and, *129*; home range, size, *126*, *127*, **128, 128, 132**, 133, 212; home range, social organization and, 156, **157**; home range, travel distances and, *130*, **132, 133, 135**, *135*; invertebrate prey, 70, 71, 72, 74–75; litter size, *176*; nest safety, 207; nest structure, *104*, 105; nesting sites, 97, 97–98, *104*, 104; nests, 97, 98; population density, *141*, 141; reproduction, output of individual females, **182**, 182; rest periods, *112*, 115; sleeping sites, *102*; social organization, 162, *150*, 156; social organization, home range and, 156, **157**; territoriality, 156, 157; travel distances, *130*; travel distances, home range size and, *130*, **132, 133, 135**, *135*; travel speeds, *130*

Tupaia nicobarica, 177

Tupaia tana (Large treeshrew), 2, 9, 23; active period duration, 111, *112*, 113, 119; active period duration, gender differences, *112*, 113, 119–20; alarm calls, 203; arboreal leaping, 45; biomass, *141*; birth seasonality, 180; body mass, 21; body mass, female vs. male, **187**, 189; body mass, home range size and, *129*, 211, 212; body mass, seasonal patterns, **187**, 189; body measurements, *20*; competition for fruit, 81–82, 83; competition for invertebrates, 84–85, 86; digestive tract, 12, 13; feeding behavior, temporal patterns, 62; fiber-spitting behavior, 60; foraging, association with birds, 166; foraging, heights for, 42–43, 44; foraging, for invertebrates, 68–69, 80, 85, 215; foraging, on stream banks, 138–40, 139; frugivory, 71, 79; fruit transit time, 59, 59; fruit trees, as specific destinations, 136, 166; fruit-eating behavior, 56; growth rates, 184, 187, 188; habitat elevational limits, 47, 47; habitat height distribution, 39, 42–43; habitat substrate use, 43, 44; home range, body mass and, 129, 211, 212; home range, gender differences, 129, 165; home range, size, 126, 127, 127, 128, 128, 132, 212; home range, social organization and, 156–61, 158, 159; home range, travel distances and, 132, 133, 135, 135; invertebrate prey, 70, 71, 72, 75, 215; juveniles, emergence from nest, 173; litter size, 176; in logged vs. unlogged forest, 49; longevity, 208; maternal care, absentee, 169–70, 195–96; maternal care, emergence of young from nest, 173, 195–96; maternal care, male time spent with mother, 173–75, 174; maternal care, nursing behavior, 169–70, 172, 208; maternal care, nursing visits, 171–73; maternal care, protection of nursing tree, 175; maternal care, teaching juveniles to follow, 173; morphology, 79; nest safety, 207; nesting sites, 98–101, 99, 103, 104, 104; nests, 100, 101; nests, structure, 104, 105, 106; nurslings, 171–72, 172; obesity, in captivity, 121; odor of, 198–99, 208; population density, 141; population density, territory size and, 159, 161; reproduction, age at first, 188; reproduction, output of individual females, 181–82, 182, 194; reproductive activity, 177, 178, 179; rest periods, 112, 115; resting locations, 113; skull, 11; sleeping sites, 101, 102, 102; social organization, 156–61; social organization, male time spent with nursing mother, 173–75, 174; social organization, territories, 163; survivorship of individuals studied, 191, 192; territorial behaviors, 161, 175; territorial boundaries, 165; territorial overlap with other treeshrews, 48; territoriality, gender and, 156–61, 158, 159, 165; territory size, population density and, 159, 161; travel distances, 130; travel distances, home range size and, 132, 133, 135, 135; travel speeds, 130; in tree plantations, 50–51, 215; vision, 14

Tupaiidae (family), 1; subfamilies of, 7

Tupaiinae (subfamily), 7, 219–21, 225; ancestry, 225; diurnality, 219; genera of, 7; insectivory, 221

Urogale everetti (Philippine treeshrew): nesting sites, 91–92, 104

Vibrissae, 17

Vision, 2, 14–15; in foraging for invertebrates, 66

Visual behavior, 15

Vocalization, 170, 172, 173; alarm calls, 175, 202, 203–4; alarms calls, resemblance to birds' calls, 204

Wade, P., 179, 181

Young treeshrews. *See* Juveniles

Text:	10/13 Sabon
Display:	Sabon
Composition:	Integrated Composition Systems
Printing and binding:	Thomson-Shore, Inc.
Index:	Jeanne Moody